AF786454

MIX
Papier aus verantwortungsvollen Quellen
Paper from responsible sources
FSC® C105338

Felix Kruse

Oil Politics

The West and its desire for energy security since 1950

Anchor Academic Publishing

Kruse, Felix: Oil Politics: The West and its desire for energy security since 1950
Hamburg, Anchor Academic Publishing 2013

Buch-ISBN: 978-3-95489-146-7
PDF-eBook-ISBN: 978-3-95489-646-2
Druck/Herstellung: Anchor Academic Publishing, Hamburg, 2013

Bibliografische Information der Deutschen Nationalbibliothek:
Die Deutsche Nationalbibliothek verzeichnet diese Publikation in der Deutschen Nationalbibliografie; detaillierte bibliografische Daten sind im Internet über http://dnb.d-nb.de abrufbar.

Bibliographical Information of the German National Library:
The German National Library lists this publication in the German National Bibliography. Detailed bibliographic data can be found at: http://dnb.d-nb.de

All rights reserved. This publication may not be reproduced, stored in a retrieval system or transmitted, in any form or by any means, electronic, mechanical, photocopying, recording or otherwise, without the prior permission of the publishers.

Das Werk einschließlich aller seiner Teile ist urheberrechtlich geschützt. Jede Verwertung außerhalb der Grenzen des Urheberrechtsgesetzes ist ohne Zustimmung des Verlages unzulässig und strafbar. Dies gilt insbesondere für Vervielfältigungen, Übersetzungen, Mikroverfilmungen und die Einspeicherung und Bearbeitung in elektronischen Systemen.

Die Wiedergabe von Gebrauchsnamen, Handelsnamen, Warenbezeichnungen usw. in diesem Werk berechtigt auch ohne besondere Kennzeichnung nicht zu der Annahme, dass solche Namen im Sinne der Warenzeichen- und Markenschutz-Gesetzgebung als frei zu betrachten wären und daher von jedermann benutzt werden dürften.

Die Informationen in diesem Werk wurden mit Sorgfalt erarbeitet. Dennoch können Fehler nicht vollständig ausgeschlossen werden und die Diplomica Verlag GmbH, die Autoren oder Übersetzer übernehmen keine juristische Verantwortung oder irgendeine Haftung für evtl. verbliebene fehlerhafte Angaben und deren Folgen.

Alle Rechte vorbehalten

© Anchor Academic Publishing, Imprint der Diplomica Verlag GmbH
Hermannstal 119k, 22119 Hamburg
http://www.diplomica-verlag.de, Hamburg 2013
Printed in Germany

Acknowledgements

I would like to show my appreciation to all the people who supported me throughout this project. Especially, I would like to dedicate this book to my family and friends, especially to Christie Brenta, who were of great help to me during the whole writing process and who therefore made my book possible.

Abstract

This book is going to analyse the International Political Economy of Oil. In order to do so, it is going to describe the impact and the relevance of oil in our everyday life. Furthermore, it is going to outline the major political incidences that took place in the period between the early 1950s and today, which can be connected to oil. This book is going to analyse these incidences through the theoretical perspective of IPE and IR realism as well as liberalism from a 'Western' point of view. It will hereby show that from the early 1950s until today the 'West' was/is following in mot cases a strongly realist approach when it comes the precious resource we call oil. This book will illustrate cases from the overthrow of the Iranian regime in 1953 until the 'race' for resources in the Arctic today.

Table of contents

List of Abbreviations

AIOC	Anglo-Iranian Oil Company (later named BP)
ARCO	Atlantic Refining and Richfield Petroleum (later acquired by BP)
BP	British Petroleum today Beyond Petroleum
bpd	barrels per day
CFP	Compagnie Française des Pétroles
CIA	Central Intelligence Agency
IEA	International Energy Agency
IOC	International Oil Company
IPE	International Political Economy
IR	International Relations
mbpd	million barrels per day
NOC	National Oil Company
UEA	United Arab Emirates
US	United States
USA	United States of America
USSR	Union of Soviet Socialist Republics
UK	United Kingdome
WMD	Weapons of Mass Destruction
WW	World War

List of Illustrations

1 Introduction

This book is going to evaluate the 'International Political Economy of Oil' with a special focus on the period after WWII. It is especially going to focus on the 'West' and its relations with oil exporting countries and is neither going to evaluate the relation between the different oil exporting countries, such as OPEC itself nor the consequences of the Chinese oil policy, the same accounts for the latest developments of the shale gas/oil revolution in the United States (US) and its potential impact on oil politics. Furthermore, in contrast to the majority of literature, which is dealing with oil politics, this book is not simply going to illustrate the major political incidents that are connected to oil after the end of WWII, but it is going to explain the reasons that made the West react in specific ways, according to IPE and IR theories.

This book will begin by describing why oil is such an essential factor for our everyday life, as well as its strategic value. Following this, this book is going to have a brief look at the 'Petro State', in order for the reader to be able to understand both sides of oil politics (the consumer's as well as the producer's side). Before this book is going to describe and evaluate the importance of state power for oil, by giving examples from pipeline constructions and the protection of water ways by the 'West', it will briefly point out the fundamental assumptions of IR and IPE theories, which this book is going to use in order to evaluate the behaviour of the 'West' in oil politics. The theories, which will be given emphasis on are IR and IPE realism as well as IR and IPE liberalism. After describing the necessity of state power for oil, this book will have a closer look at the events, which took place in the Middle East in the early 1950s. The evaluation will include the developments before 1973, including the overthrow of the Iranian government in 1953, as well as the disempowerment of the 'Western' oil companies by the newly founded Organisation of Oil Producing Countries (OPEC). In a second paragraph, the 1973 oil crisis and its aftermath will be evaluated, including the reaction of the International Oil Companies (IOCs) to the crisis, as well as the reaction of the 'Western' governments. In addition, this book is going to describe the US relations with Iran and Saudi Arabia after 1973 by mainly focusing on the US attempts to stabilise the oil price, including the second oil crisis. This will be followed by a description of the tasks of 'International Energy Agency' and the

‘International Energy Forum’ and an evaluation of their work from an IPE/IR perspective. Following this, this book is going to evaluate the US import restrictions for oil in the middle of the last century and will also assess the structure of the oil market from a theoretical perspective. Lastly before the final summary of its findings in the conclusion, the book is going to have a brief look at the latest developments currently taking place in the Arctic from an IR and IPE perspective.

2 Prologue

Since the beginning of the twentieth century, oil has been entwined with security, power and the position of nations (Yergin 2011: 126).
The security of energy is not only about countering a vast number of threats; it is even more about the relations among nations, how these states interact with each other, and how the security of energy has an impact on their overall national security (Yergin 2011: 264-265).

In 1911, Winston Churchill decided to alter the firing of the British navy from coal to oil. This decision, which might seem on the surface quite technically and irrelevant for this book, was going to change the twentieth century fundamentally. By switching from coal to oil, the British Navy was no longer relying on a home-based form of energy for its fleet, which was Welsh coal, but on oil. From that stage on, the Navy had to rely on the distant and insecure oil supplies from Persia. However, the benefits of this source of energy outweighed the disadvantages by far. One century after Churchill's decision, oil is still central to the security, prosperity and the very nature of civilisation. Since the days of Churchill, his words have remained as present as in 1913 when he was saying that: 'the safety and certainty in oil was lying in variety of sources and variety alone' (Yergin 2008: xiii- xiv; Yergin 2011: 265).

Until some alternative source of energy will be found in sufficient scale, oil will keep its far-reaching effects on the global economy; a low oil price is like tax cuts for the economies of oil importing countries. Paying less for home-heating and gasoline has the effect that the consumer has additional money, with which domestic economic growth can be stimulated. Furthermore, a low price is an antidote to inflation, which allows countries to grow faster, with lower interest rates and a minimised risk of inflation and vice versa. Additionally, oil stays a massive generator of wealth, not only for individuals and companies but also for entire nations, as it has been shown in the past. Furthermore, mainly due to the fact that it is co-moving negatively with the US dollars exchange rate, oil became also an important financial commodity[1] and cheap supply of energy for the home country became a big topic in national elections.

---

[1] Oil as a financial commodity please see appendix

However, oil is not a commodity comparable to cars, planes and other goods. Oil is in most cases always interwoven with 'national strategies', global politics and above all power (Yergin 2008: xiii- xvii; Yergin 2011:106). Whoever is in the position to control the outflow of oil is and was in the position to starve his enemies of this vital resource (Parra 2010: 1).

Furthermore, oil became the driving force of economic growth after WWII leaving coal far behind. Today without oil there would practically be no mobility, and without energy in the form of oil to create electricity there would be no 'Internet age' (Yergin 2011: 264-266).

However, some advocates may argue that the upcoming of other sources of energy might make the discussion about oil (gas and coal) less relevant. However, even if sources such as nuclear technology, solar power, as well as wind were combined they would not be able to replace the energy, which is currently generated from oil. For example in order to replace oil completely by nuclear energy one would have to build an additional 4,000 1,5 giga-watt nuclear power stations (today the world counts 440). These new reactors would deplete all known uranium reserves within ten years (Downey 2009: 26-27).[2] In addition, the accident in Fukushima threw a new question mark over the 'global renaissance' of nuclear technology, which may have solved some issues in the rising global demand for energy.

To sum up, there is one thing one can be certain about; that in the year and decades to come the world appetite for energy and especially for oil will enormously increase and among it so will the political struggle (Yergin 2011: 2-8).

---

[2] numbers refer to 2009

3 The strategic importance of oil

The history of oil since the time of Churchill is strongly intertwined with security issues. From the 'western' point of view, oil has been found in all the wrong places.[3] The access was in most cases difficult and always far away. Additionally, oil was found almost exclusively in less developed nations, which became increasingly difficult to deal with as they developed politically and economically mostly due to oil revenues (Parra 2010: 293; Chalabi 2010: 2).

3.1 WWI and WWII

In 1914, European nations began the first partly mechanised war. The used vehicles in this conflict, such as planes and especially cars and trucks were more efficient than anybody could have predicted and they pressed into large-scale military service. If it had not been for the 'taxi fleet' of Paris, which brought tens of thousand's of troops to the front, Germany would have most likely conquered huge parts of France if not even defeated it. As a consequence of the strong involvement of vehicles in the war (more then 160.000) in 1918 the US had the highest ever-recorded gasoline price in their history until today, in inflation adjusted terms. After the end of the war the strategic importance of oil became imminent to the world leaders. From that stage on the security of supply became their major strategic objective. This idea is one of the major reasons why three former Ottoman Turkish provinces, which were expected to be rich in oil, were bound together and formed the country today known as Iraq (Yergin 2011: 229-230; Engdahl 2004: 35-44; Yergin 2008: 152-156).

During WWII, oil proved once again its vital significance from a military point of view. During the conflict, the Allies used seven billion barrels of gasoline to win the war, out of which, 85 per cent came form the US. During the conflict, oil proved to be the key to many different issues. Furthermore, it was the main reason for Japan to attack the US after it put de facto an oil embargo on the country. Japan's fear that a lack of oil would turn its battle ships into nothing more than scarecrows was one if its core arguments to go to war and had a huge influence on the strategy it used. By attacking Pearl Harbour, Japan wanted to

[3] except for the USA until the end of WWII

ensure the safety of its oil supply routes, which stretched from the Philippines until the Japanese mainland (see illustration below).

Japanese oil supply routes during WW II

(Map made according to Yergin 2008: 342)

In Europe, Hitler Germany decided to invade Russia not only for ideological reasons, but also with the aim to get a hold on the Russian Caucasus during Operation Blue/'Fall Blau'. It was also planned that General Erwin Rommel would fight his way through Africa and the Middle East to unite with the other German troops in the Caucasus (please see illustration on the following page).

Operation Blue/'Fall Blau' illustration

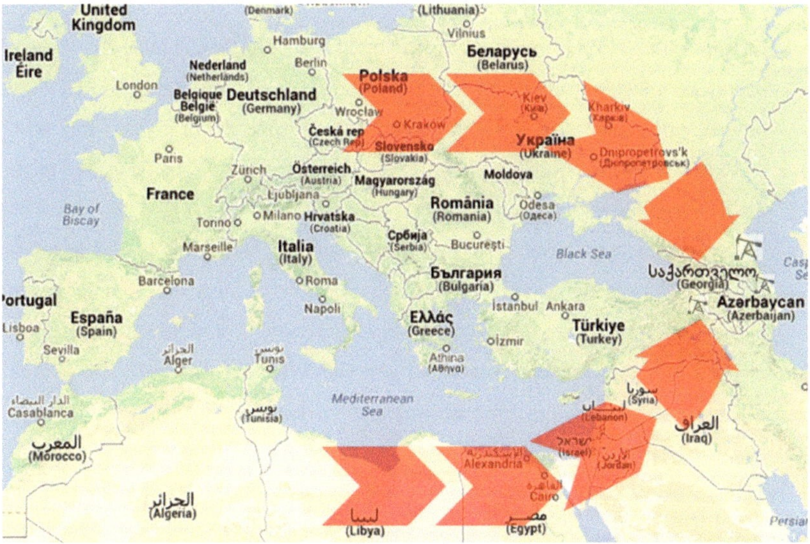

(Map made according to Yergin 2008: 321)

However, a lack of supply in oil set an end to General Erwin Rommel's Africa campaign, as well as to the operation 'Barbarossa' in Russia and to General George Patton's sweep through Europe after the landing in the Normandy.

During the war, the interruption of the Allied supply lines was a prime target for German submarines, which were attacking the tankers of the Allies crossing the Atlantic (and almost succeeded). The Allies followed the same strategy and tried to cut off Japan and Germany from their oil supply more successfully- in the Pacific with submarine attacks on Japanese tanker and in Europe by bombing German factories for synthetic fuels.[4] This led to such a shortage that planes were pulled to the runway by farm animals or did not have enough fuel to start at all. In Japan the shortage in conventional fuel led to the manufacturing of what we call today bio fuels made of potatoes, rice, sugar etc.. After the war had ended the German army had only 11,000t of gasoline left. The situation in Japan was similar. Both countries started a war without oil and lost it to a major extent due to the lack of it. The strategic significance of oil became once again apparent (Yergin 2011: 230-231; Aust and Richter 2011; Yergin 2008: 300-350).

[4] Germany's only source of fuel after 1944 (Yergin 2008: 300-350)

3.2 The first Gulf war

On August 2nd 1990, the Iraqi dictator Saddam Hussein ordered his army to invade his neighbouring country Kuwait (Yergin 2008: xiv). His main aim was not primarily to destroy a neighbouring country but its oil reserves, while other motives were at best secondary. However, the story is not as simple as one might think. In 1990, Iraq had acquired a massive debt due to its war with Iran. This debt was to a large extent financed by Kuwait and Saudi Arabia and Iraq was fighting for its economic survival after the end of the conflict with Iran. However, the low oil price at that time made it very hard for Iraq to reduce its debt and to recover from the long and destructive war with Iran. The reason for the low oil price was seen from the Iraqi government as the result of an overproduction, especially by Kuwait and Abu Dhabi above OPEC quotas. Consequently, the Iraqi government accused the governments of Kuwait and UAE of waging economic warfare against Iraq. [5]

However, when the tension between the Kuwait and Iraq increased, the USA ambassador in Iraq gave the fatal signal that the US would not intervene in internal Arab affairs, which was seen as a green light by Iraq to invade its neighbouring emirate (Parra 2010: 295-296). The reward for controlling the small country was enormous. The successful and permanent invasion of Kuwait would have meant that Iraq would have become one of the worlds leading oil power by controlling about 10 per cent of the global oil reserves at that time (Parra 2010: 298, 305; Yergin 2011: xiv). [6]

With its increased resources Iraq would have rivalled Saudi Arabia as the dominant oil power, with a far reaching impact on the Persian Gulf, where the vast majority (2/3) of the world's 'conventional' oil reserves were located at that time. The impact for the rest of the world would have been consequently far reaching as well (Yergin 2011: 10). The power that Iraq would have gained from its new position would have forced the world to pay court to the ambitions of the Iraqi dictator, which would not only have had the oil reserves, but also vast

[5] Saddam Hussein saw the USA as the main agitator behind it (Parra 2010: 295-296). Additionally, the Iraqi government accused Kuwait to illegally exploit Iraqi oil reserves, by drilling horizontally from Kuwaiti territory for oil on Iraqi territory (Parra 2010: 295-296).

[6] In addition, a major part of its debt from the war against Iran was held by the Emirate, which was gone too, as a consequence' of the invasion (Parra 2010: 298, 305; Yergin 2011: xiv).

amounts of revenues from oil trade and the fourth largest army by numbers in the world. The consequence would have been a historic shift in the international balance of power (Yergin 2008: xiv: Yergin 2011: 10). However, an intervention was not crystal clear from the beginning, as one might assume, as one adviser to the US president said: 'We will have to get used to a Kuwait-less word'. But the picture was more alarming then the adviser had guessed. If Iraq could not be removed from Kuwait, how long would it take before it would attack Saudi Arabia and the UAE? These countries would be under a constant threat. This scenario would have had totally unacceptable consequences for Israel and the price of oil.

However, before the fighting started the 'western' coalition estimated its losses of around 10,000 to 15,000 soldiers[7], which shows the huge risk it was willing to take to free Kuwait or better to say the risk it was willing to take to protect Saudi Arabia (Parra 2010: 300-303). [8;9] Following this, the war aims were simpler than officially proclaimed: the first aim was to eliminate the threat of Iraq to the UAE and the Saudi Arabian oil reserves, while the second one was to leave Iraq strong enough, in order for it to remain a political unit to counter balance Iran. After the war had ended, the presence of US forces in the Middle East seemed to have appeased the region and to have secured the supply of oil (Yergin 2011: 14; Parra 2010: 302, 311-312).

3.3 The second Gulf War

The reasons for the second Gulf War were vastly different than the ones of the first one. The so called 'coalition of the willing', was willing to go to war with Iraq in 2003, contrary to 'Old Europe'. The reasons for the war were primarily not interlinked with oil at all. The primary factors were: the attacks of September 11, 2001 and its consequences, the threat of WMD, the way that the first war in 1991

---

[7] Actual casualties were 150 during the operation on the coalitions side, Iraqi casualties are estimated between 85,000 and 200,000 (Parra 2010: 301).
[8] By comparison during operation 'Overlord' on the June 6th 1944 (D-Day), the allied troupes claimed 10,000 casualties and 4,414 deaths (ddaymuseum.co.uk).
[9] Referring to this the Joint Chief of Staff Powell and General Schwarzkopf said the following: 'I think we'd go to war over Saudi Arabia, but I doubt we'd go to war over Kuwait' (Powell), 'I don't see us going to war over Kuwait. Saudi Arabia, yes, if we had to; but not Kuwait' (Schwarzkopf) (Parra 2010: 300-303).

had ended, including the persistence of Saddam Hussein's regime and the way the intelligence services had analysed their data.

After Saddam Hussein was captured and interrogated, he replied to the question, why he had kept the illusion of having WMD – Iran.

Oil did not play the major reason when it came to the Iraq war in 2003. The significance of oil was simply highlighted by the nature of the region; the vast amount of oil in the Middle East and the critical balance of power in the region. However, the potential danger from Iraq in 2003 was not as significant for the region as it had been more then ten years earlier. Furthermore, in 2003, neither the British nor the US government were pursuing a mercantilist 1920s-style ambition to control the Iraqi oil. The real issue was not who owned the oil at the well, but whether it was available for the world market.[10] It was the idea that a democratic Iraq would become a more reliable supplier of oil and that it could quickly increase its oil production, since the UN sanctions on the country would be abolished.[11] However, the war was not over as quickly as expected and it cost the US taxpayer a trillion dollars in direct outlays until 2011[12] (Yergin 2011: 141-149; Klare 2004: 94-105; Chalabi 2010: ch.15).

[10] Oil from Iraq could be purchased on the world market but only to a limited extent and under UN restrictions (Yergin 2011: 141-149; Klare 2004: 94-105).

[11] A sustainable increase in the Iraqi oil production after the war compared to 2003 did not materialise until 2010 (Fiedler 2012).

[12] this amount equals roughly the nominal income from oil revenues by all OPEC countries in 2011 (US Energy Information Administration 2012).

The other side of the coin: The petro-state

For oil exporting countries a low oil price has opposite effects than the positive ones for oil importing counties, which were described in the prologue. In petro-states oil revenues account for 50 to 90 per cent of the government's revenues. These 'easy' revenues have a huge influence on the states and the society's structure, the psychology and the motivation of people. In most cases, innovation, entrepreneurship, hard work and the development of a competitive oriented economic growth are the 'casualties' of these state systems. The economies of petro-states become inflexible and are loosing the ability to adapt and change. Furthermore, the state apparatus controls most of the economic life from economic growth via micromanagement and grand projects, etc. Additionally, there are two more characteristics that refer to countries, which are rich in resources. One is called the 'Dutch disease', named after the negative effects of the 'new' gas wealth the Netherlands acquired in the 1960s. As a result of the gas wealth, the national currency became overvalued and exports became relatively more expensive on the world market. This led to the decline of the non-gas related export economy. Furthermore, local businesses started becoming less and less competitive due to an increase in cheaper imports and an increasingly embedded inflation. All these negative effects became known as the 'Dutch disease'.

The second, even more debilitating illness of the petro-state is a seemingly irrepressible fiscal rigidity, which leads in most cases to excessive government spending irrespective of the macroeconomic situation and became known as 'the reverses Midas touch' (Yergin 2011: 107-111).[13]

[13] For more details about the 'reversed Midas touch' please see appendix.

Theoretical background

International Political Economy and International Relations developed several theories in order to explain states behaviour. The following paragraph is going to outline the core theoretical assumptions of two IR theories (realism and liberalism) and two IPE theories (Economic nationalism and Economic liberalism). These four approaches are the most common and are able to explain most developments taking place in the field of IPE or IR. However, the arguments supporting why certain developments take place differ according to the perspective one follows.

Theories, such as Marxism, Feminism, Neoliberalism, Constructivism and so on are not thoroughly investigated in this chapter, but are worth mentioning.

3.4 Classical Realism

The tradition of classical realism can be tracked back almost 2,500 years ago to Thucydides and his writings about the Peloponnesian War. However, Niccolò Machiavelli, Carl von Clausewitz and Hans Morgenthau were also some of the most influential scholars in this school, alongside Thucydides.

In the domestic sphere, realists argue that the state is a guarantor for law and order. However, in the international sphere there is no power above the state level. Therefore, realists see the world as an anarchic place, where each state is looking for opportunities to take advantage of one another. Therefore the survival of the state, which is its highest objective, depends on its material capabilities and its alliance with other states, in order to establish a balance of power. In order to survive, each state is trying to gather as much power and unilateral advantage as possible, whereby power has to be always seen relatively in comparison to a or the closest rival and does not only need to be perceived as military power. However, the successful exercise of power requires a sophisticated understanding of the goals, weaknesses and strengths of the allies, opponents, as well as third parties. Furthermore, the struggle for power is seen as a zero sum game, which means that the increase of power of one state is the loss of another. This struggle is seen as eternal, due to the unchanging human nature.

Realists tend to regard history as cyclical, in a sense that states try to build order and escape from an anarchic world. States might be successful in doing so, however, only for a certain period of time, which is always followed by anarchy again.

Many IR scholars agree with Morgenthau's six principles, which are especially pointing out the following:

The first principle claims that objective laws, which have their roots in human nature, govern society as well as politics. The second one points out that states are exclusively thinking in terms of power. The third principle argues that the motive of power-seeking is an objective category and has universal validity. The forth principle claims that the self-interest of a state is a higher good than socially constructed moral beliefs. Morality cannot be applied to the state only for the individual. The fifth principle assumes that every state has 'its own' moral beliefs, which do not have universal validity. The sixth principle points out that realism cannot be judged by the methods of other sciences (Morgenthau 1978: 4-15; Jackson and Sørensen 1999: 76-80; Lebow 2010: 59-68).

Realists also claim that international organisations are mainly shaped by powerful states, according to their will and for their own purposes (O`Brien and Williams 2010: 21- 24; Russett 2010: 110-112).

3.5 Economic nationalism / Mercantilism

In general realist IPE is regarded as a subset of IR realism. According to the assumptions of Thucydides, who saw wealth as a critical source of military strength by pointing out that war would be a matter not so much of arms but of money, which made arms of use (Cohn 2010: 56-57).[14] However, IPE realism is sharing the basic assumptions of the realist approach from IR, such as the states struggle for power, prestige, and influence. Additionally, it defines wealth as another issue for the struggle between states. Both theories see world politics as a never-ending conflict, aiming to pursue more power, whereby power is measured vis-à-vis other states. Since the participation in the market is potentially seen as

[14] Or as Paul Kennedy put it: 'The historical record suggests that there is a very clear connection in the lung run between an individual Great Powers economic rise and fall and its growth and decline as an important (military) power' (1988: xxii).

negative, economic nationalists are arguing in favour of state control of key economic activities or for state assistance to vital economic sectors.

Therefore, IPE realism, in contrast to IR realism, also focuses on what IR realism defines as 'low' politics, which is associated with commercial and financial pre-eminence. Furthermore, IPE realists turn issues in the world of economic affairs into problems of international economic diplomacy and trade wars. This means that nations are solely concerned with their own interest when they engage into international economic negotiations. In this context, economic nationalism is focusing on the importance of power when it comes to the sharpening of the international political economy, in which the state is seen as the main actor. Furthermore, IPE realists argue that the nature of the global political economy reflects only the interests of the most powerful states. Economic nationalists also believe that there is only a limited amount of wealth in the world and that each state has to block the interests of other states in order to protect its power/wealth in a zero-sum game. International organisations are, from this perspective, seen as having their capabilities delegated to them by the state(s), and are therefore relying on them. Furthermore, international institutions are seen as arenas for acting out power relationships, in which the most powerful nations 'shape' the rules to 'fit' their interest. IPE realism also sees the state as prior to the market and market relations are designed by political power. Nevertheless, it also recognises the importance of actors in the market, such as the role of companies. However, mercantilism subordinates their importance below the level of the nation. In the end, this theory argues that firms are always subject to the dictate of the state and that they are generally viewed with suspicion (O'Brien and Williams 2010: 22, Cohn 2010: 56-57, Ravenhill 2008: 32-34).

3.6 IR liberalism

The most important competitor for the realist approach is the liberal approach, which can be tracked back to Immanuel Kant, John Locke and Hugo Grotius and is seen as an especially 'western' theory.

Liberals believe that nations despite their self-interest are capable of cooperation and have the ability to create a harmonious and peaceful society.

Liberals assume that conflicts and wars can be mitigated or overcome, through concerted changes in the domestic as well as the international structure of governance. Scholars, such as Kant, see democratic governments, international law and institutions, as well as economic interdependence, as means to overcome the security dilemma of the international system. The just described 'Kantian triangle' is founded on the following assumptions. It is claimed that international organisations (such as the UN or the WTO) are playing a major role as mediators and supervisors between different nations and are stimulating peace. It is also claimed that economically important trade between nations creates incentives to maintain peaceful solutions with each other due to interdependence. Furthermore, trade, and especially free trade, is seen as a win-win situation, from which all participants can benefit from. Additionally, free trade is seen as positive due to the fact that it is preventing wars, which are motivated by the need for resources. Additionally, it is claimed that democratic countries will refrain from using force against each other and are creating a 'liberal zone of peace', due to the fact that the elites of a country are accountable for their actions to the people, who generally want peace. Hence the domestic political system is determining a nation's outside behaviour, which also puts the individual, as well as interests groups into the focus of IR liberals (Russett 2010: 96-110; O`Brien and Williams 2010: 21-24).

3.7 Economic liberalism

Liberal IPE is sharing the basic assumptions of IR liberals and is in most cases just using different arguments/a different perspective (mostly economic arguments/an economic perspective) to point out the same assumptions. Nevertheless, liberalism has the highest influence on the world's economic system, as we know it today (Cohn 2010: 77).

For liberal scholars IPE is constituted by the search for wealth. Contrary to economic nationalism, liberalism is focusing more on the individual or a wider range of actors from the state, over companies to interest groups. The state is just seen as one of the actors, which are creating the international economy. Furthermore, liberals assume that people and states can cooperate for their mutual benefit and that interaction between entities results in a positive sum game, creating a win-win situation. Hence, liberals are always searching for fields and conditions for cooperation and assume that conflicts are evitable through interdependence. Liberals such as Adam Smith and David Ricardo, are also arguing that free trade is the most beneficial and productive economic system, maximising growth and wealth through comparative advantage and division of labour. Individuals and companies are seen as the key economic actors and as the main source of wealth. Many liberals view the state with hostility and vote for non-interference, since it would bring politics into the field of economics and therefore into the market, which is the centre of economic life. However, most liberals agree on the fact that the market has imperfections, but do not agree on which measures should be taken. Transnational companies are also seen as positive for the home as well as for the host country.

Liberals also argue that economic nationalism, with its protectionist policies, is leading to conflict and that only a liberal approach is beneficial for peace. Liberals also support the idea of self-determination and the use of international organisations for the settling of disputes between countries (O'Brien and Williams 2010: 21-25; 41-45; Ravenhill 2008: 41-47).

With IR liberals, IPE liberals share the assumption of the pluralist nature of the international system and the feasibility of cooperation and interdependence. However, liberal IPE does not consider democracy as an essential factor for its assumptions, since it is primarily focusing on macroeconomic arguments, as Friedrich August von Hayek among others pointed out (Daily Bell, unknown year).

The importance of state power for Oil

For more then one century, oil has been essential to modern economies, while imports have become a matter of economic survival (Parra 2010: 293).
The usual definition of energy security is very simple: the availability of sufficient supplies at reasonable prices. However, there are several dimensions when it comes to the security of oil. Starting from the physical security, which includes the protection of all assets; infrastructure, supply chains, trade routes as well as making provisions for a quick replacement and substitution if necessary. Furthermore, the access to oil itself is crucial as far as the ability that needs to be acquired and the supplies that need to be developed physically, commercially and contractually are concerned. Additionally, energy security is also a system that consists of national policies as well as international institutions, which are designed to respond in a coordinated way to emergencies, dislocations, disruptions as well as helping to uphold the steady flow of supplies, such as the International Energy Agency (IEA) (Yergin 2011: 265-268).

Finally, there is the constant need for investment on behalf of the state or private companies. Therefore, the security of energy requires policies and a business climate, which encourages investment and development to ensure an adequate supply infrastructure in the future. However, for oil exporting nations the question is 'slightly' different. These countries are thinking in terms of the security of demand. They are relying on the perspective that there will be a market in the future as a basis on which they can plan investments and are able to plan their budgets (Yergin 2011: 266-267; Yergin 2011: ch. 5).

3.8 The Baku–Tbilisi–Ceyhan Pipeline

Pipelines have the huge advantage of being built within a long time perspective, regarding the delivery of a certain quantity of oil to a specific place and customer.
After the collapse of the Soviet Union, 'new' independent sovereign nations came to existence around the Caspian See, which were rich in oil. The development of these resources was inextricably entangled with the ambitions of nations and of

course geopolitics (Yergins 2011: 43). However, when it comes to natural resources there are many players in the game, as it usually happens.

The Caspian Sea is not connected to any other ocean or sea. The richest in oil and key state in the region is Azerbaijan on its west shore (please see map on the following page). Azerbaijan is bordering Armenia and Georgia to the West and Russia to the North. In the South of the country is Iran, which has ambitions to be the dominant power in the region. All these states and of course the United States, Britain, Russia and China, as well as Turkey had a dominant role in reshaping the region. The major point of conflict for all these states was on which route and how the Caspian 'gold' would be exported to the world market (Yergin 2011: 43-52).

Map of the Middle East

(geology.com: unknown year)

All the participants in the game had different ambitions. Russia wanted to keep its influence on the countries and called it until the mid 1990s 'our oil/their oil'. The 'West', represented primarily by the only left super power USA along with Great

Britain, saw the great chance of diversifying the world markets supply and by doing so, increasing the global energy security. In order to do so, their aim was to prevent Iran from filling the vacuum Russia had left and to prevent the new states from sliding back under exclusive Russian influence.

Iran and Turkey both saw their chances to increase their influence upon the former Soviet Republics. Hereby Azerbaijan was of particular importance to the Iranians due to the fact that Iran has a 'minority' of 16 million Azerbaijanis[15], who were connected to the Azerbaijan population in many cases through family bonds. An open tolerant 'western' regime in the country could have challenged the Iranian regime due to the close connection of this ethnic group.
China was also interested in the region during the 1990s due to its rising need for energy imports and its proximity to central Asia.

The objective for the 'Caspian' nations was clear, i.e. maintain and consolidate their independence and establish themselves as nation states (Yergin 2011: 47-49; Mammadyarov 2007).[16] The development of the oil resources could not only bring strongly needed revenues, but also political support and interest in the countries. Oil was as an Azeri put it their strategy, defence and independence. However, when it came to the development of these countries not only politics were essential but also pure economics, such as engineering costs, investment, logistics etc.

After oil companies from seven nations acquired exploration rights (among them Russian, US, Japanese and Europeans) the main question was how to bring the oil to the market. This question was not only of interest for the companies but even more pressing from a political perspective. There were several possible routes to follow, in order to export the crude oil to the world market.

The Russian government wanted to connect the future pipeline with its own pipeline infrastructure, which would have given Russia some degree of leverage and control over the Caspian oil. However, the United States strongly opposed this plan and pushed on every occasion possible for the route that was built eventually. Another option would have been to go through Georgia and to the Black Sea. The Russian and Georgian option both would have included a

[15] Iran has a total population of approx. 78 million (CIA World Fact Book 2012).
[16] Mammadyarov has been Foreign Affairs Minister of Azerbaijan since 2004.

transport via tanker over the Black See and through the Bosporus. However, these two routes were problematic due to the fact that the narrow Bosporus was already on its limits concerning the shipping traffic.

Nevertheless, there was another route, which would have been the cheapest, fastest and easiest route to construct. In this scenario Azerbaijan would have delivered oil to Iranian refineries in the north of Iran, and Iran would have exported via a swap the same amount/value of oil through its terminals in the Persian Gulf. However, this cheap alternative would not have benefited any government except for the Iranian. It would have given Iran the control over Azerbaijan. Additionally, it would have increased the reliance on the Gulf and would have run contrary to any diversification strategy of the 'West'. Additionally, it would have given the regime in Tehran additional revenues, which would also have been opposing 'Western' interests.

The last, most expensive and complicated option, which was eventually realised, was the option to build a pipeline from Azerbaijan towards Georgia, circling around Armenia[17] and leading through Turkey to the Mediterranean port of Ceyhan (please see map on the following page). This route was finally constructed[18] due to the high pressure of the 'West' especially the USA and Turkey, despite the fact that Russia obviously opposed it (Yergin 2011: 55-60).[19]

[17] a rivalling state to Azerbaijan at that time (Yergin 2011: 55-58)

[18] Later on a gas pipeline followed the same route as the oil pipeline (Yergin 2011: 55-58).

[19] Two smaller pipelines had been built prior to the main pipeline in 1996 but only in order to get revenues to the country asap and to level the interests between the USA and Russia. One pipeline went through Russia and the other one through Georgia, both were ending at ports at the Black See. The pipeline through Georgia was also important, because the transition fees were helping to stabilise the country and it avoided the conflict area of Chechnya (Yergin 2011: 55-58).

Caspian Region Oil Pipelines

(Peak Oil Optimist 2005)

Today this pipeline is transporting more than 1 per cent of world's oil supply or more than 1 mbd. During the Georgian – Russian war in 2008, Russia fired more than 50 missiles on the pipeline trying to destroy it. A successful attack would have meant a stronger dependence of Western Europe on Russia, which would have been the only possible provider of infrastructure for the oil export of Azerbaijan (McElroy 2008; Brauer 2001).

This example illustrates, how IPE/IR realist's approaches to world politics influenced the building of this specific pipeline and pipelines in general.

From a liberal point of view (IR and especially IPE), it would have made sense to build the pipeline through Russia or even Iran. Both liberal schools would have argued that by building the aforementioned pipeline, one would create interdependence between the 'West' and the transit countries. The transit countries would have welcomed and/or needed the revenues for their households and the 'West' would have needed the oil. These facts would have made a pipeline through both countries very attractive for liberals. IPE liberals would have even favoured the cheaper Iranian version, since it would have been the cheapest and fasted one and according to the 'rules of the market' this would have been the solution to support. However, one has to be cautious with IR liberalism because Russia and Iran are/were no democratic states, which is the only form of government IR liberals support, since democracy is essential for their assumptions in order to fulfil the 'Kantian triangle'.

Realists (IR and IPE) however, evidently differed strongly from the liberal standpoint. Realists would argue that a pipeline through Iran or Russia would have given the states more power, since they were in the position to block the supply of oil to the market. Consequently, concerning these states, the 'West' would have lost, on the one hand some degree of power for the benefit of Iran or Russia. On the other hand, by building the pipeline through Georgia one would not only have brought Azerbaijan closer to the west, but also Georgia due to the money the pipeline would have poured into the government's pockets, which would have meant additional power and influence for the 'West' in this region.
IPE realists in this context, contrary to IR realists, would also have taken into consideration the loss of power compared to the reduction in wealth the most expensive route would have cost compared to the Russian and the Iranian route. However, the Iranian route would have been unquestionably out of consideration to any alternative costs for obvious reasons considering the relationship between the USA and the Iranian regime. The Russian solution might seem reasonable from today's point of view.

Nevertheless, the Cold War had just ended and the future development of the country was not so clear at that time, hence that solution was off the table as well.[20]

The only other solution IPE realists could have agreed with, in contrast to IR realists, would have been to take the rout via Georgia and the Bosporus and not via Turkey, considering the minimised costs. However, since reduced costs would have been beneficial for Azerbaijan only in form of a higher surplus from selling the oil and it would not have been directly beneficial for the 'West' itself, there was consequently no reason not to build the safest pipeline.

[20] Or as Madeleine Albright put it at that time: 'We don't want to wake up ten years from now and have all of us ask ourselves why in the world we made a mistake and didn't build that pipeline (the safe one via Turkey) (Yergin 2011: 60). This is strongly reflecting the realist assumption that relatively peaceful times are always followed by conflict, since history is seen as cyclical.

3.9 Water ways and state power

As history has shown, energy security does not only need to be understood as the security of energy supply itself, but moreover as the protection of the entire supply chain, from the initial production to the final consumer (Yergin 2011: 280). The vulnerabilities of the existing infrastructure take many forms, from outright hostile assault to small events that are able to trigger a massive disruption of the energy market. Very critical in this context are choke points at see routs for the transportation of energy in form of military conflicts, accidents or terrorist attacks.

About 50% of the world's oil production is transported through these choke points. These choke points of interest in general and in particular are, the Strait of Hormuz (over 23 mil. bpd.), the Malacca Strait (40 mil bpd, 80 per cent of Japan's and Koreas supply and 40 per cent of Chinas supply), the Bosporus Strait (over 3 mil. bpd), the Bab el-Mandeb Strait (up to 3 mil. bpd), the Suez Chanel (up to 2 mil. bpd) and the Panama Chanel (0.6 mill. bpd). All major navy forces are trying to keep these routes safe and to protect them especially from piracy, such as in the case of the Horn of Africa (Somalia). However, of all these choke points the biggest presence of naval forces is represented by the US, such as the presence of the US Firth Fleet in Bahrain in the Persian Gulf shows, where 60 per cent of the conventional oil reserves are located. For the protection of its oil supplies on these waterways the US are currently spending about $ 50 billion per year. [21] However, this presence does not exist without some bad feelings from states such as China, which on the one hand profits from the protection, but also fears to be cut off from supply in the case of confrontation (Yergin 2011: 282-284, 293, 303; Aust and Richter 2011; YI 2009; globalsecurity.org 2012; González 2012).[22]

---

[21] According to a speech of US Secretary of Defence Robert Gates in 2010 the US Navy is as strong as the following 13 navies combined (Holmes and Yoshihara 2010).

[22] The correlation of these choke points and the presence of the US navy can be seen on the two maps on the following page.

World Oil Choke Points

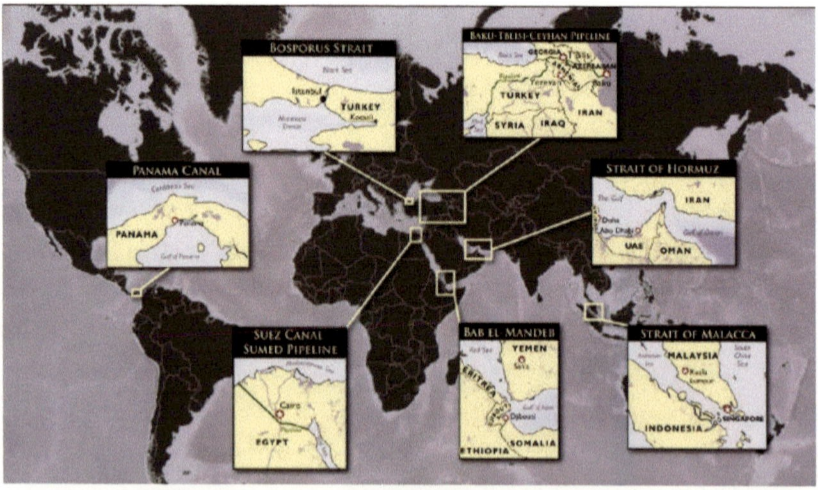

(The Lugar Energy Initiative, unknown year)

US Navy Bases and Facilities in 2012

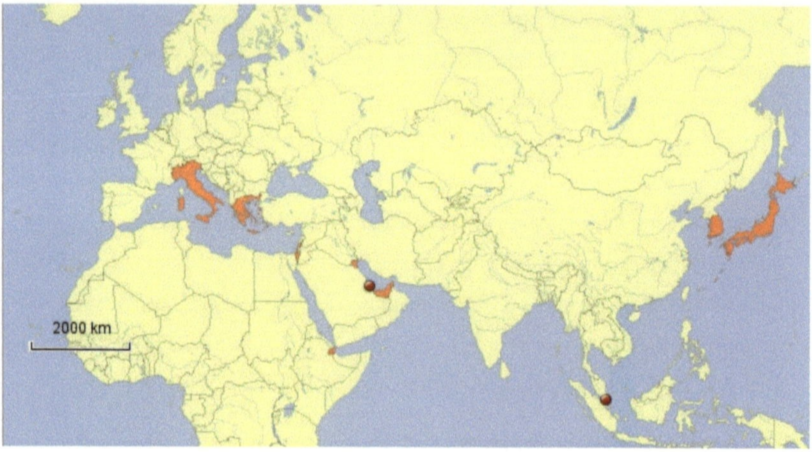

(Map created according to: NavyAdvancement.com)

This example of the concentration of the US/'Western' navies along these choke points is another illuminating example for the realist approach the West is following in the context of oil.

Liberal IPE and IR scholars could, on the one hand, argue why it is necessary that the consumer/recipient is protecting the product until its destination and not the producer as it is normally the case, which would save the 'West' and especially the US a fortune.[23] On the other hand, realists see in the control of the sea routes a very strong tool of power since it is enabling the 'West'/US to starve every enemy of this vital source. IPE realists would argue in the same way, but would evaluate the costs of the huge fleet against the increase of power. They might come to the conclusion that it is not worth spending that much money on waterways and that it might be wiser just to protect certain ones instead of all of them. However, to let any potential rival nation have control of these points would be out of question for IPE realists, as well.[24]

---

[23] Of course the costs for the protection of the waterways would be include in the price in this case. However, all consumers would pay it in that case not manly only the US.

[24] A solution for example would be, just to control choke points, which are not only related to oil but to global shipping traffic as well, like the Strait of Malacca or the Suez Chanel and to encourage the oil exporting states to protect the Horn of Africa for example in a cooperation with the 'West' or on their own.

The oil price, the Middle East and international oil companies

When the security of energy is under discussion, the analysis always returns to the Persian Gulf, which accounts today for more than 60 per cent of the so called 'conventional' oil reserves (Yergin 2011: 4-5).

However, in the following chapter, this book will analyse the major political incidents before and after the first oil crisis in 1973. Nevertheless, when one is looking at the policies of the 'West' in connection with the Middle East or oil in general, one has to keep in mind that the political circumstances have changed over time and that different legislations and nations followed sometimes different approaches in this context.

Furthermore, the political actions that took place during the 1970s and beginning 1980s always have to be understood under the assumption of the governments of producing and importing countries that oil would become short soon. It was also expected that the Soviet Union would become a net importer of oil shortly. In addition, one has to keep in mind that until 1970 apart from Russia, Canada and the US, next to the OPEC countries, there was no major producer of oil in the world (China was a minor one) (Parra 2010: 6, 114-117; Chalabi 2010: 127).[25]

3.10 The pre 1973 developments

Since Harry Truman (1945- 1953), U.S. presidents have made the security of the Middle East, and especially the Kingdom of Saudi Arabia and its natural resources a fundamental national interest. Especially when the US turned from an oil exporting country to an importing country in the beginning 1950. Jimmy Carter said in this context that any attempt by outside forces to gain control of the Persian Gulf region, would be seen as an assault on the vital interests of the United States and would be replied by any means necessary, including military force. This threat against an outside aggressor has to be seen in the context that in 1979 the Soviet Union invaded Afghanistan, which was understood as an attempt

[25] For a diagram of the Geopolitical incidents from 1960 till 2008 and the oil price please see appendix.

to expand Russian influence into the Gulf region.[26] However, in turn of the American guarantee of the safety of the region and especially of the Kingdom of Saudi Arabia, the country tied its long-term security strategy to the US (Yergin 2011: 126- 127).

However, by 1950 all oil companies that were active on a global scale, especially in the Middle East, were of US and Western European origin. Most of these Middle East countries had and still have in some sense a 'client relationship' with the USA or Britain.[27]

The relationship between the host countries of the oil companies, the home countries of these corporations and the oil companies themselves has been and still is until these days in the heart of policies and politics in the international petroleum industry (Parra 2010: 1-2; Chalabi 2010: 2).

The legal basis on which the oil companies worked in the host-countries, have been established well before 1950. Despite the fact that the concessions of the companies were all individual, they had certain points in common.

The oil companies could for example extract as much oil as they wanted for a certain period of time. The companies had to make payments as compensation for their extraction, which included surface taxes, royalties, production taxes and so on and so forth (Parra 2010:6-9; Chalabi 2010: 2).

In the beginning of the 1950s, these concessions changed quite dramatically for the first time. Starting from Venezuela, a 50/50 deal worked its way through the Middle East to split the profit of the posted price[28] 'equally' between the International Oil Companies (IOC) and the host countries. Prior to these changes the host countries sometimes received less then 10 %.

However, the international oil companies started, prior to the nationalisation, to keep the posted price artificially low and to 'take' the margin between the 'real' price (market price/value) and the posted price. This development finally was one of the major reasons why OPEC[29] was founded in 1960 and why the nationalisa-

[26] This Russian aggression was successfully fought back by insurgents financed and equipped by the USA with the help of Saudi Arabia, where most of the insurgents came from such as Osama bin Laden (Bronson 2006: 8- 10).

[27] The military expending for the protection of the oil resources in the Persian Gulf accumulated according to 'The National Defence Council Foundation' to 52-62 billion p.a. in 2003 (Copulos 2003:ch. 2). For a map of the client state please see appendix.

[28] For a detailed explanation of the 'posted price' please see appendix.

[29] For a list of all current OPEC member states please see appendix.

tion of the oil industry in the Middle East eventually took place. At the beginning OPEC had the primary objective to guarantee a stable and high price for oil by representing a 'united front' against the IOCs, in order to increase the bargaining power of the counties.

Additionally, the host countries realised in the early 1950s that the income taxes the IOCs were paying them were being credited by the origin country of the companies. Therefore, they did not directly affect the companies themselves, but only the origin countries of the companies. This fact was also realised by the US State Department, which was favouring the idea of higher revenues for the oil exporting nations in the Middle East and concluded that higher payments would be desirable and could increase the stability of the local governments. The fact that these additional revenues came into the region without the necessity of any parliamentary approval put further 'icing on the cake' (Parra 2010: 15-20; Chalabi 2010: 33, 35).

However, before the foundation of OPEC and the nationalisation of the oil industry took place at the end of the 1960s/beginning 1970s, there was another major incident, which took place in the early 1950s.

In 1951, Iran passed a law and started to nationalise its oil industry. The following 2 years were characterised by the negotiations between the British, US and the Iranian government and AIOC (Anglo Iranian Oil Company)[30], as well as the World Bank, the Security Council and the International Court of Justice. In August 1953, the Iranian government was overthrown and a new government was formed under Prime Minister Fazallah Zahedi.

However, the motives behind the overthrow and the politics that followed were only to a minor extent directly related to oil, as there was more than enough of it in the Middle East at that time. Nevertheless, it nicely illustrates how the 'West' promoted its interests in this part of the world in that specific period.
However, towards the end of 1952 Britain and the United States had accepted the reality of nationalisation and only the marketing of Iranian oil and the compensation for the IOCs were still a subject of discussion.

Nevertheless, ultimately things developed differently. Due to the fact that without AIOC Iran was unable to sell its oil to the world market, it had a huge loss

[30] later called BP

of revenues in the years of 1952 and 1953. As a consequence, the government of Mossadegh, the Iranian leader, appeared more and more volatile and possibly vulnerable to a left-wing, pro Soviet takeover, especially to the US government. For the British government the loss of an important source of sterling oil was the centre of their concerns apart from the Soviet threat. Operation AJAX was launched by the CIA only due to the communist threat, in order to replace the Iranian government of Mossadegh in 1953 (Parra 2010: 22-26, 27; Downey 2009: 11; Engdahl 2004: 95-97).

However, during the time of nationalisation the output of oil in Iran almost dropped to nil, whereas the average Middle East production grew by 12% between 1951 and 1953. After the Shah came back into office, AIOC did not get its monopoly back in the country, but found it self in a consortium with Shell (14% share) and the five US major (Chevron, Exxon, Gulf, Mobil and Texaco (each 8% share) and CFP (8% share). In order to improve the situation of Iran, the US and British government put pressure on the oil companies to 'buy' more oil from Iran. However, since the market was controlled by almost one 100% by these companies, they simply reduced the production in other Gulf countries and increased it in Iran, which kept the prices stable.[31] This behaviour became afterwards a common procedure for 'Western' governments, in order to finance its partners in the Middle East, without the need for a debate in their houses of political decision-making[32] (Parra 2010: 22-30, 46, 71-72, 103).

However, the defeat of Iran by the 'West' after the nationalisation in 1951 made the members of OPEC play quite moderately after it was founded (till the beginning 1970s), not looking for any confrontation. Furthermore, as long as the Shah was in office, the 'West' tried to use him to have a pro 'Western' influence on OPEC as well as Saudi Arabia in some occasions (Parra 2010: 64, 89-96, 103-107). But the Shah had another even more important role. After the withdrawal of the British forces from the East of Suez, he was chosen by the US to fill the power vacuum Britain had left, in order to protect the sphere against any Soviet agitation

[31] For a detailed illustration of the ownership links between major IOCs and major crude-oil producing companies in the Middle East please see appendix.

[32] Usually, Iraq was the country suffering most from this practice because its exports were in most cases reduced in favour of others (Parra 2010: 22-30, 46).

and influence.[33] Additionally, the Shah was a useful tool for the 'West' because he was quite moderate towards Israel (Parra 138-140).

However, the establishment of OPEC in 1960 became a significant step for its member states. It was the first time that they had some leverage against the IOC and were able to demand higher revenues or increased shares without the threat that the IOC would play one country off against another. During that period (1960-1973), when the IOC were forced to pay more and more to the host country for their oil production or were forced to give up more and more shares of their production companies in the Middle East to the local governments, the IOCs did not get any backing by the 'Western' governments against these developments. The 'West' was just always demanding terms that were reasonable and affordable for both sides.

The USA and Western European countries neither wanted to move alone or, which accounts especially to the US, did not want any confrontation with the OPEC governments, especially the Shah, and only urged for moderation in demands. However, the main aim of the US government was always to avoid the cut off of the supplies. Nevertheless, the US government of course knew that it was not paying the bill for the price increases, due to the fact that the majority of the Middle Eastern oil would go to Europe, since the US had import limitations and generally a higher oil price then Europe (Parra 2010: 128-139).

In short, the consuming countries wanted assured supply at reasonable prices, with emphasis on the 'assured' and not on the price. The only thing they really cared about was the future availability and the security of supply, nothing else. This of course was also realised by the OPEC governments, which therefore used the disruption of supply as their main threat in the Arab Israel conflict, in order to reduce the support of the US for Israel (Parra 2010: 137, 147, 167).

However, since the Middle Eastern nations were outright petro states, they wanted since the beginning of the 1960s, more and more control over their main industry, which included assured markets for their petroleum at reasonable prices to sustain their governmental spending's. At the end of the 1960s/beginning 1970s, the governments could no longer stand the situation, as they had to rely on foreign companies for their revenues, which could have abruptly changed when

[33] the main aim of the US policy at that time

the companies would have decided to shift more export to another country, hereby putting the companies interest over the nations. Furthermore, most OPEC countries felt betrayed by the IOC, which were earning the huge differential between the agreed posted price and the market price (Parra 2010: 137, 146; Chalabi 2010: ch. 5).

As the tensions between the IOC and the OPEC states got too high, the nationalisation followed. During the nationalisation of the oil industry the 'West' only argued that the compensation was too low. However, there was no resistance that is worth mentioning (Parra 2010: 159, 168, 210).

So, how can we judge the just described events in the Middle East from a theoretical point of view?

The overthrow of the Iranian government can be clearly seen as an IR/IPE realist manoeuvre, with the aim to reduce the Soviet influence/power in the region. The loss of the mostly populated country in the region to the 'enemy' would have meant a huge loss of power to the 'West'. Furthermore, having the Shah in the long run as policeman in the region, which would be depending on the support of 'Western' governments in order to stay in power, gave the 'Western' governments a huge leverage. Liberals (IPE/IR) would not have intervened in the processes in Iran respecting the right of self-determination of the people and would only have tried to persuade the government to liberalise the market and initiate democratic reforms and maybe would have assured financial support to get the reforms going. Furthermore, liberals and especially IR liberals, with their focus on democracy would have by no means installed and supported the Shah as a more or less autocratic ruler. IPE liberals would of course, always claim that they demand democracy. However, history proved a different outcome, such as the support of the Pinochet dictatorship in Chile (1973- 1990) by Freidman and Hayek, two leading IPE liberals in the 20th century illustrates. In this context Hayek said the following during his visit in Chile: 'Personally I prefer a liberal dictator to a democratic government lacking liberalism' (Daily Bell unknown year; Friedman 1975). Therefore it can be argued that IPE liberals would have supported a regime with a liberal economic policy. However, an overthrow of a regime in general would have been unlikely.

When one is evaluating the reaction to the nationalisation of the oil industry, one can be 100 per cent certain that neither realists nor liberals, from a 'Western' perspective, were favouring this move. Realists would be primarily concerned about the loss of power with respect to the OPEC states, which were no longer directly relying on 'Western' companies.[34] As a consequence, the 'Western' governments lost the ability to directly influence the households of the OPEC countries. Liberals, which are favouring free trade, as well as a free market and see transnational corporations generally as positive (especially IPE liberals), would of course argue against the nationalisation as well, since it is against their beliefs in a free market. However, whether liberals would really argue against the nationalisation, is at least questionable, since the ownership of the production was a relict of colonial/past-colonial times and was not the result of a fair and competitive market. Furthermore, the nationalisation was made according to international law.

However, it would not have made any sense for the oil importing nations from the perspective of any theory to suddenly develop a hostile, aggressive policy towards the OPEC countries and thus against countries, on which they believed to depend on for by far the biggest part of the increased oil supplies that the importers would need over the next decades.[35] Additionally, the fear of the oil weapon was already very real in the minds of 'Western' politicians (Parra 2010: 167-170).

Having a look at the increases of royalties that the companies had to pay to the host country prior to the nationalisation, the arguments are different to what we have seen before from a theoretical point of view. IPE realists from the US would have strongly appreciated an increased price for Middle Eastern oil, since it was reducing Western Europe's competitive advantage in respect of cheap energy and vice versa. IR realists would not have considered this phenomenon at all, since it was not directly influencing the power relations between the countries. Therefore, it would have been considered as an issue of 'low' politics. Liberals (especially IPE liberals) would of course always argue that the price is not set at the market

[34] Of course the states were still relying on the IOCs since those companies were owning the whole distribution network, starting with the tanker fleets and ending with the gas stations in the 'Western world'.

[35] The US for example expected to be relying in 1985 on more then 12mbd. from OPEC countries, which was more than half of the expected US consumption in that specific year (Parra 2010: 167-170).

and is therefore 'artificial' and that the market should be liberalised and set the price.

3.11 The 1973 oil crisis and its aftermath

On October 6 1973, the day of Yom Kippur[36], Syria and Egypt launched a synchronised surprise attack on Israeli forces in the Golan and the Sinai Heights. Israel repelled the attack and regained lost ground after the US made the decision to supply the Israelis heavily with military equipment (U.S. Department of State-Office of the Historian, unknown year).
On 17 October 1973, ten OPEC countries met in Kuwait and announced that they would reduce their oil production by 5 per cent each month until Israel forces would evacuate all occupied territories. Despite this announcement, the US announced a massive new military aid program for Israel on 19 October. The following day, Saudi Arabia proclaimed a ban on shipments of oil to the US.
However, the oil weapon was a dangerous weapon to use for the Arab nations. The Gulf producers, who were cutting back most of the supply for the market were all military 'midgets' compared to the consumer nations. Additionally, to some point, withholding supplies from the industrial nations, especially the US, would become a fighting matter as Kissinger and Ford pointed out several times.[37] This threat of military force was replied by Saudi Arabia by declaring that it would destroy its oil wells in the case of an attack (Chalabi 2010: 1; Parra 2010: 180-185; Engdahl 2004: 137).

However, since the embargo was only directed against a few states, it was easily circumvented by reshuffling the suppliers. As a result the embargo only led to a rise in transportation costs, which were more or less equally spread on all customers because they all were in some sort of oil-sharing program (Parra 2010: 184-185). Furthermore, the international oil companies tried their best to counter the embargo by shifting the supplies for the different countries in a way that countries with an embargo against them would get non-embargo oil from other countries. Hereby it is important to point out that no oil company favoured its

[36] the holiest religious day for Jews
[37] In this context is has to be kept in mind, that at that time the Nixon administration was still engaged in the fatal Vietnam war, and was trying to withdraw its troupes without damaging the US credibility/ reputations (BBC News unknown year).

home country when it came to the distribution of oil, as an investigation of the US Federal Energy Administration showed, even when governments tried to pressure them as the British did for example (Parra 2010: 188, 275).

However, even though the embargo did not materialise at the end, it had quite a big impact on the market and on international politics. The disruption the embargo created led to panic, chaos, shortages and economic confusion all around the world. This materialised in a scramble between companies, traders and countries for oil. Additionally, the shock led to a massive change in geopolitics and shifted the economic balance of power away from the 'global north' to the 'global south'. For the 'Western' countries themselves, the situation was leading to an enormous strain and antagonism, as they struggled to respond and started to blame each other and tried to outmanoeuvre each other, in order to secure their supplies (Yergin 2011: 269-274, 280). As a consequence of the embargo governments tried to make more government-to-government sales, in order to secure the future supply after 1973. (Parra 2010: 185-186). All these experiences directly led to the establishment of the IAE in 1974.

However, only the higher oil price in the aftermath of the embargo made the exploitation of oil from the North Sea economically reasonable, with the very positive effect of diversifying the sources. Additionally, it reduced the competitive advantage of non-US countries that had better access to cheap Middle Eastern oil (Parra 2010: 143; Engdahl 2004: 136-137; Chalabi 2010: 3-7).

An evaluation: The support of Israel is almost impossible to be explained from a realist perspective.[38] It was very clear to the 'West' that the support of Israel would endanger the supply of the vital source of oil and therefore played against the national interest of the 'Western' nations concerning their power, wealth and their economic strength. Additionally, it had the potential of pushing some Arab nations further to the huge rival of the cold war the USSR. Realists would have claimed that the support of Israel would be motivated by ideology, which is seen as fatal for guiding a nation's foreign policy.

Liberals, especially IR liberals, would have doubtlessly supported the only democracy in the Middle East against the aggression of its autocratic neigh-

[38] One could argue that it was the moral obligation of the US to support Israel and that by doing so, it gained prestige and more power among its 'Western' Allis and strengthened the 'Western' alliance.

bours.[39] Furthermore, one has to bare in mind the influential lobby groups the Israeli state had/has in the US. This fact was reflected, from a liberal point of view, in the US foreign policy during that time.

The threat of military force by the 'West' in order to break the embargo can only be explained from a realist perspective. It was the national interest of the 'West' to have unlimited access to the vital source- oil. Without this vital source (oil), the West could not survive. Therefore, from a realist perspective, it was necessary to threat the Arab countries with the use of its hard power, in order to guarantee the access to this vital source for the economy and the nations survival. Liberals however, wouldn't have used the threat of force in order to regain access to the oil from the Middle East. For liberals the only way would have been a diplomatic one, trying to convince the Arab nations to return to (free) trade with the 'West'.

When it comes to the mutual outmanoeuvring of the 'Western' governments, in order to secure as much oil for themselves as possible, it is again only explainable with a realist approach, where the survival of the own nation is the only aim, regardless of other nations. It might seem amoral for a liberal, which would have searched for a joint solution with all oil importing states. However, realists do not act in categories such as moral or amoral. For realists it was just about the nations survival and nothing else.

3.12 The US relations with Iran and Saudi Arabia after 1973

In the aftermath of the oil embargo prices stayed very high. One the one hand due to the fact that the Israeli- Arab conflict was still unsolved even though the war had officially ended on October 25, 1973 and on the other hand due to production cuts from OPEC countries. An additional reason was the end of the Bretton Woods system in 1971 and its consequences for the value of the US dollar (Cohn 2010: 343; Chalabi 2010: 68-69, 167).[40]

In 1974, Saudi Arabia argued against most of the other OPEC members that the oil price was already too high with a posted price of 11.651$/barrel and was

---

[39] IPE liberals as described before do not generally or necessarily consider democracies as positive. However, the economic policy of Israel is/was certainly more liberal than the one of any Arab nation and therefore IPE liberals would have also supported Israel.

[40] For a more detailed explanation please see appendix.

worried about the economies of the industrialised countries. Scholars, such as Parra, point out that there is little doubt that this position was also in response to what is regarded as the US position, which was pressuring Saudi Arabia to bring the oil price down. However, Saudi Arabia insisted that the US should also pressure the Shah to go along with the Saudi position. When the US declared to do so, but eventually did not, Saudi Arabia stopped to follow US pressure concerning the price of oil (Parra 2010: 196-203).

However, in this context, it is important to know that the US had a huge leverage/power on/over both countries; both nations were depending on US supply for their military, US financial institutions to handle their governmental funds and both were also relying on US help in the case of a Soviet attack or the attack of another predatory neighbour (as the Gulf War in 1990 demonstrated) (Parra 2010: 201-203). However, behind the scenes the US did not want to weaken the Iranian oil revenues, due to the fact that they were mostly used to buy weaponry against the Soviet threat and the US were also aiming to install Iran as the 'local policeman' of the Gulf (Parra 2010: 203; Chalabi 2010: 101).[41] Furthermore, Europe was the biggest importer of 'Gulf oil' and therefore basically paying most of the price for the armament of Iran, which bought its weapons from the USA equalling a value of almost 9 billion $ between 1972 and 1976 (approx. 30 billon as NPV 2011) (Parra 2010: 203-204).[42]

Until the end of the 1970s and therefore just in time for the second oil crisis, a spot market for oil out of 'Western' production had developed and was often referred to as the real price for oil by the masses.[43] However, the spot market was very successful in misleading some OPEC countries[44] over the real value of oil and the prices, that could be achieved (Parra 2010: 230-231).

Nevertheless, during the second oil crisis in 1979 and 1980 Saudi Arabia was selling its oil below the price of other OPEC countries and additionally increased

[41] In this context it has to be kept in mind, that the USA were at that time involved in Vietnam and did not have any domestic backing for filling the vacuum that Britain left in the area when it withdrew its troops from 'East of Suez' between 1968 and 1971 (Parra 2010: 203-204; Reynolds 2000: 216).

[42] The US bought most of their gulf oil from Saudi Arabia (Parra 2010: 230-231).

[43] The spot market itself had practically no direct influence on the flow of money. It was at that time a very small market in terms of money and volume and had therefore no real connection to the vast volumes traded by the OPEC countries (Parra 2010: 230- 231).

[44] Not to say all OPEC countries except for Saudi Arabia (Parra 2010: 230- 231).

its oil production to bring down the oil price.[45] There are several explanations to Saudi Arabia's behaviour. The political motives were mainly a reaction to the US policy in the Middle East, and especially the Saudi's need for free access to US military goods and US support in general.

Some scholars, such as Daniel Yergin and Fadhil Chalabi see the main reason for the lower oil price of Saudi Arabia in the fear of the government to damage the world economy in which the nation had a growing stake, as well as in the fear that a high oil price would motivate the 'West' to develop alternative fuels on a large scale, which would be a major threat to all oil exporting countries (Yergin 2008: 685; Parra 2010: 234-235; Chalabi 2010: ch. 12-13).[46]

---

[45] In 1979 Saudi Arabia sold its oil for $18/barrel and the OPEC countries for $28/barrel. In Dec. 1980 Saudi Arabia sold one barrel for $32 and the other OPEC countries for $36/barrel (Parra 2010: 230- 231).

[46] However, there is also another explanation by Francisco Parra a former 'Secretary General of OPEC'. He sees the reason for the behaviour of Saudi Arabia in a deep sense of insecurity. Pointing out that the only power of the country was the world price of energy, due to its oil power by producing 40% of OPEC oil. However, at the same time Saudi Arabia was military relatively weak and faced several hostile neighbours. Additionally, the country had internal tensions resulting from the division between its western and eastern provinces and tensions between the royal family and the newly rich merchant class as well as tensions between immigrants (mostly guest workers) and the native population. The only reasonable solution after the experiences from Iran in 1953 was to keep the head down and not being seen as the leader of the price hunt in order to stay out of trouble (Parra 2010: 234-236).

The theoretical explanation:

The US reaction, concerning the high oil price and the pressuring of Saudi Arabia and not Iran to bring it down, cannot be explained from an IR perspective, since IR theories are about the management of the relations between states and does not primarily imply ordinary economic issues, which are considered as 'low politics'. Liberal IPE scholars would have generally criticised the US policy in this case. IPE liberals see the intervention of the state into the market process generally as wrong and harmful to the economic life. However, since the selling price was not the result of supply and demand in the first place, liberals (IR/IPE) would have argued to liberalise the market in general.

IPE realists would have argued that this split policy of the US was in this context the attempt to achieve two aims. On the one hand, not to harm an important Ally (Iran), which was projecting US influence into the gulf and therefore power.[47;48] On the other hand pressuring Saudi Arabia to put cheap oil to the market can be understood as an attempt to reduce the average oil price to a level, which was not harmful for the economic development of the 'West' in general. Additionally, one has to keep in mind that the US mainly bought the cheaper Saudi Arabian oil, whereas Europe had also to purchase the more expensive Iranian one as well. As a consequence the competitive advantage of the US increased in energy matters. Furthermore, the 'additional' petro dollars got recycled through arms purchases in the US.

3.13 The IEA and the International Energy Forum

The foundation of the IAE was the direct result of the Arab oil embargo of 1973, which put the 'Western' alliance to the test. Since its foundation in 1974, the IEA has/had the task to coordinate the emergency sharing of supplies in the case of a major loss in supplies. However, it is the obligation of the national governments whether to implement any of these measures or not. Additionally, the IEA is nowadays trying to establish a dialogue with all OPEC and non-OPEC

[47] This attempt would have been shared by IR realists.
[48] In this context it has to be said, that the installation of an autocratic regime as 'policeman' in a certain area of the world is for liberals unthinkable.

countries, as well as with all oil importing countries, which are not in the IEA (Yergin 2011: 269-274, 280; Parra 2010: 190-191).

The International Energy Forum (IEF) is the final step in the dialogue between the oil importing and exporting countries, representing 90 per cent of the global oil and gas market. Both OPEC and IEA are members. The real test of its functionality was first in 2008 when the members were trying to restore the stability of the oil market and to reduce the price of petroleum in order to reduce the damage to the world economy. However, in 2011 during the price increase, the forum showed its limitations, which can especially be traced back in this case to the OPEC countries of Saudi Arabia versus Venezuela and Iran. It demonstrated that the dialogue is determined by the relationships not between specific 'blocs', but between certain states and to which extent they can act in respect of their national interest (Yergin 2011: 274-275).

Despite the difficulties, the IEF the IAE proved to be a very useful tool for the industrialised countries to coordinate their energy policies. The IAE in the following primarily focused/focuses on: decontrolling of crude oil prices as well as products. However, according to IAE, prices should be allowed to reach a level on which substitutes and energy saving would be encouraged. Additionally, the establishment of certain standards for the efficiency of cars and other vehicles was the aim of the IEA. Furthermore, it was/is their aim to reduce the dependency on oil imports, meaning by this term to reduce the dependency on OPEC (Parra 2010: 246-248).

The theory:

The establishment of the IEA as a supranational institution can clearly be seen from a liberal point of view (IR/IPE) as an attempt to reduce conflicts between the oil importing nations. All member states are democracies and are also bound together by trade. These three aspects can be found in the 'Kantian triangle' as a guarantor for peace. IPE realists do not necessarily see democracy as essential in this context as it has already been mentioned above. Realists (IPE/IR) however, would argue that the IEA has no real power as an institution and is relying on the power the nations are willing transfer to it. As the text above points out, the implementation of any IEA measures depends on the individual governments, which is what realists would argue against this institution.

The IEF can be interpreted similarly as a liberal attempt to reduce tensions this time between oil importing and exporting nations even though not all member states are representing democracies, which is important from a liberal IR perspective as well. However, as described above, the limitation of this institution is the national interest of its member states, which is perfectly compliable with realist's criticism of international institutions in general.

3.14 US import restrictions, the oil price and a 'competitive' market

The most important aspect of oil from the Middle East has been and still is that it is incomparably cheap to produce. Until the late 1990s, it could have easily been shipped to the US and, despite the additional costs of freight driven most of the local producers out of business. The same accounts for most other oil producing countries worldwide.[49] However, these US producers were mostly politically very powerful, rich and large in number. Following the argument that imports were threatening the national security, imports (except from Canada and Mexico) were in a first phase supressed by political pressure and later regulated (1959) via a quota system until the early 1970s (Parra 2010: 45-47, 251).[50]

In 1986, when the oil price collapsed Saudi Arabia, it fixed the oil price to 18$/barrel (which was a rise in prices). This was the result of a complex political-economic deal of disparate elements, which also included Iran, the US and Libya.[51] It is only known that there was a deal, what it exactly consisted of besides the sales of weaponry to Saudi Arabia such as AWACS and Tornado airplanes could not be retraced (Parra 2010: 288, 322; Chalabi 2010: 250, 267).[52] Iran was depending on oil revenues in order to finance its war against Iraq (which was supported by Kuwait and Saudi Arabia). However, the most important factor

[49] As a matter of fact the Middle East could have supplied the whole world demand in 1999 (if it had been so minded and therefore developed enough capacity) by charging only 5 $ per barrel and still earning sufficient revenues for their governmental needs (Parra 2010: 45-47).

[50] The import restrictions were of course also helpful for the balance of trade in the long run. However, weather cheaper energy would have given the home economy a bigger competitive advantage in a macroeconomic context can also be discussed at this point.

[51] Besides the US also Great Britain put pressure on the governments of the Gulf states especially Saudi Arabia (Chalabi 2010: 209).

[52] It took Saudi Arabia some critical manoeuvres to turn the market over the year but at the end it succeeded (Parra 2010: 288, 322).

were the US producers, which could only survive when the oil price was around 20$ as the government had calculated. A similar intervention in order to restore the oil price took place by the US with the help of Saudi Arabia after the price collapse during the Asian financial crisis in 1998 (Chalabi 2010: 250, 267; Parra 2010: 288-291).

However, for oil importing nations (represented mostly by the IEA and the United States) it might seem desirable to achieve security of supply by encouraging competition between the oil exporters, so that any attempt to exert political pressure by withholding supplies would lead to the loss of markets to a competitor. Nevertheless, the risks of this course have been seen as way too high to follow, due to the fact that the Middle East oil is incompatibly 'cheap' and that it could easily drive other high-cost sources out of the market, which would lead to an even stronger dependence on this region. An agreement of all importing nations to keep the prices high via tariffs seems unrealistic in this context. Additionally, the non-OPEC producers are obviously well protected by the OPEC price umbrella, which is basically guaranteed by Saudi Arabia as the only swinging supplier[53] of oil and are therefore not criticising the system. (Parra 2010: 294, 335-336; Yergin 2011: 287).[54]

However, this strategy is not without any consequences for oil importing nations, which have in general a negative trade balance with oil exporting countries, which is getting 'artificially' bigger due to the higher price.[55]

The four perspectives:

IR Liberals see protectionist measures as the cause for conflict since it is on the one hand, hindering the development of interdependence and can, on the other hand, lead to conflicts about resources. Even if the latter is not the case for US oil since there is more than enough outside the US, it is still minimising the aspect of interdependence. IR Realists however, would have strongly favoured import restrictions and the protection of the domestic oil industry for its strategic value, making the US less dependent on the international oil markets (exporting coun-

[53] For a more detailed description of OPEC and its role as a swinging supplier please see appendix.
[54] According to an analysis of Sanford C. Bernstein from 2011 the worlds leading 50 oil companies are having an averaged production cost per barrel of approx. $90. For Saudi Arabia this number is significantly smaller at around $45 per barrel (Peaple 2012; Downey 2009: 69).
[55] For a detailed explanation please see appendix.

tries) and therefore vulnerable. Since the import restrictions ran out in the early 1970s the 'manipulated' market took the same function.

Opposite to IR realists, IPE realists would have carefully evaluated the costs and benefits of the import restrictions and the artificially high oil price, since the costs of energy imports account for the biggest chunk of the US trade deficit, since the beginning 1990s. One could argue that this deficit is also becoming a major risk for the security of countries, since it is making states more dependent on the international monetary markets, which can have fatal results as the 'Euro crisis' shows at the moment. However, it seems that this risk is seen as less significant than the risk of getting cut off the supplies. In this case, IPE realists would have also argued that the current system is only reflecting the interests of the most powerful states (the 'West'), which can only be perceived as correct considering the above-mentioned facts.

The strongest opposition to a regulated market is, of course, coming from IPE liberals. They would argue that the artificially high oil price is preventing the market to work, according to their theory. This means that since there is no free trade and no free market where the 'invisible hand' is setting the price for goods countries are not optimising their production in respect of their comparative advantage/resource abundance. This leads to a suboptimal use of resources and a reduction in the division of labour. IPE liberals may also claim that only a very small percentage of the population is benefitting from the artificially high prices, which are, in this case, the producers and the consumers, the majority has to pay the price. Similarly to IR liberals, IPE liberals would also claim that through the reduction of interdependence conflicts become more likely.

The consequences of an 'Oil-rush' in the arctic. On the edge of a 'new Cold War'?

A topic that has been drawing continuous media attention is the possible 'oil rush' in the Arctic.[56] Along with this discussion goes the debate about a 'new Cold War' motivated by the scramble of nations for the precious resources located in the Arctic. However, the question remains whether these fears are really founded on solid grounds or are just a 'parody' of Cold War times (Johnston 210).[57] Most attention has been drawn on maritime boundary disputes between Canada, Denmark, US, Norway and Russia. However, according to Danish estimates, 95% of Arctic resources are within internationally recognised national borders, as the following map illustrates.

Map of the Arctic Region

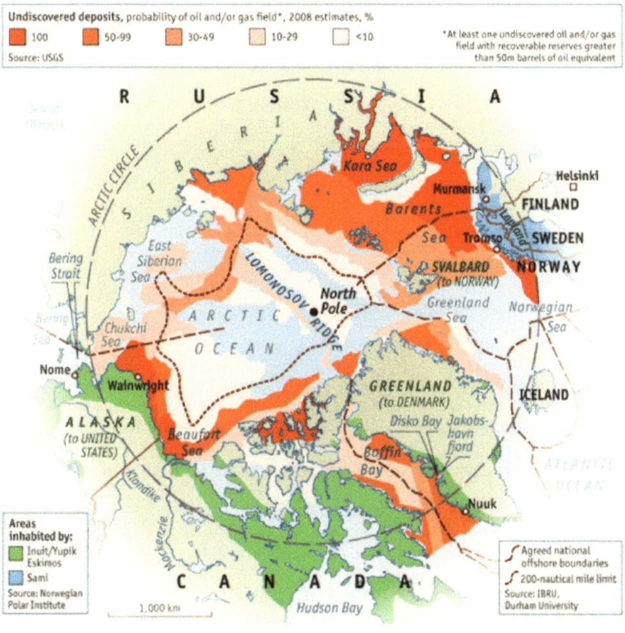

(Source: Economist 2012)

---

[56] and even more a gas rush

[57] By comparison, Saudi Arabia is having estimated reserves of about 264.1 billion barrels.

Furthermore, there are supplementary reasons, which make a peaceful solution about the disputed areas very likely.

First of all, there is the profit motive. The Arctic states will generally first develop the undisputed resources they own and then argue about those they do not have (Economist 2012: 10-13). An indicator for this can be the agreement signed between Russia and Norway in 2010, which resolves one small border dispute between both countries in the Varangerfjord, which is part of the Arctic. A similar agreement has been reached between Iceland and Norway, which decided to develop the Dreki field in a joint venture (Jonston 2010). Another very strong reason for the Arctic states to cooperate and to solve any conflict peacefully is the reluctance among those states to give outsiders any excuse to intervene with regional affairs (Economist 2012: 10-13). This is illustrated by the fact that all states have directly or indirectly agreed to the United Nations Convention on the Law of the Sea as well as the 1958 Convention on the Continental Shelf to solve any border dispute.[58] Furthermore, the Nordic States have set up the 'Arctic Council', which is dealing with Arctic issues (Johnston 2010).[59] Another motive for cooperation are the high costs of operation in the huge Artic region. For this reason, all members of the 'Arctic Council' signed a binding agreement in 2011, to co-ordinate search-and-rescue efforts.

However, the highest/only potential for a military conflict between the Arctic states is between NATO states and Russia. If Russia would start to enforce its claims militarily it has no chance that this action will pay off, since the NATO capacities are much higher in general and in the region especially. Furthermore, Russia has neither the technology nor the know-how to develop the oil by itself. Hence it is in the interest of the Russian NOCs to enter into cooperation with 'Western' oil companies to exploit their potential resources. However, since these 'Western' companies will only do so in a peaceful environment companies, such as Lukoil etc. use their strong influence on Moscow to counterbalance the military attempts or any nationalist escalation (Golts 2011: 52-59; Laruelle 2011: 75-84). An additional indicator for a peaceful solution is that the defence chiefs of all

[58] Only the US has not ratified the Convention. However, it is expected that this will happen soon. Furthermore, the US joined the other Arctic states in issuing the 'Ilulissat Declaration' in May 2008, which is claiming that the law of the sea is the overreaching framework to solve any dispute (Johnston 2010).
[59] For additional information on the 'Arctic Council' please see appendix.

Arctic states have met in 2012, in order to discuss defence co-operations between them, a meeting, which is supposed to become an annual event. Furthermore, there have been discussions about an Arctic oil spill response between all states (Economist 2012: 10-13).

Moreover, the changes in the military structure and equipment – for example, the strengthening of the Canadian Rangers or the move of the main Norwegian land unites to the north of the country - have only little to do with an attempt of power projection into the areas of the Arctic with uncertain ownership. They primarily have the purpose of patrolling and protecting the recognised national territories that are becoming more and more accessible, including illegal activities (Wezeman 2012: 13-15).

However, the above-mentioned developments can be interpreted from different perspectives. Realists would claim that the peaceful developments in the Arctic only take place, since the power of the NATO is too big to be counter balanced by Russia. As a consequence, if Russia wants to be treated equally it is forced to go along with the installed institutions and agreements, which according to this theory are mainly shaped by the interest of the most powerful nations. IPE realists would add to these aspects that an arms race in the Arctic would/could outweigh the economical benefits, which is another reason for a peaceful solution.

Liberals would argue from a different angle. Liberals, and especially IPE liberals would argue that the peaceful developments are caused by interdependence and the installed international organisations and agreements. The interdependence is forced through the need of 'Western' technology and know-how to develop the Russian oil resources. Hence the NOCs counterbalance the military solution in the scramble for the resources, as described above. This underlines the liberal assumption that the domestic political circumstances determine the outside behaviour of nations. On the one hand, the 'West' has an interest in the diversification of supply from an area, which is seen as peaceful/reliable compared to the Middle East. On the other hand, the 'West' has a strong interest in a financially stable Russia, which is currently depending on the exploitation of its natural wealth. The international organisations and agreements are seen as the foundation for this peaceful solution. IR liberals would share all these assumptions of interdependence and international organisation, however, since Russia is no spotless Democracy, not all aspects of the 'Kantian triangle' are fulfilled.

Conclusion

In this book we have seen that without oil our everyday life, as we know it, would not exist. It has also been discussed, which crucial influence the oil price has on the importing as well as the exporting nations economies and therefore the nations themself. The strategic value of oil was proven during two World Wars and when we are looking at the contemporary uprising in Syria it seems that a lack in petroleum is becoming a crucial factor for the Assad regime in loosing control over the country. However, the first and second Gulf War proved once more the willingness of the 'West' to fight for oil security, oil price stability and the free access to this vital source for the world economy.

Chapter six proved that the 'West' is pursuing a strongly realist approach when it comes to the security of their oil supply routes and that in this case, economic view points and the interest of private companies are (at best) secondary. The developments before the first oil crisis also gave evidence about the realist policy the 'West' followed/follows when it came/comes to oil and its security. This fact was demonstrated by the treatment of Iran and the Shah by the 'West' during the Shahs ruling, as well as by the reaction of the 'Western' governments to the nationalisation of the oil industry in the Middle East in the early 1970s. The first and probably only suspension of this realist policy in this context could be seen during the Yom Kippur war where the 'West' and especially the US supported Israel against the aggression by its Arab neighbours, which was running contrary to almost any realist approach. As a consequence of the oil embargo, one could also witness that most 'Western' nations quickly put the well-being of their own nation first and that there was barely anything that could be called a 'united front' from the 'West' opposing the embargo; the opposite was the case.

By assessing the US-Iranian and US-Saudi relations after the first oil crisis we were able to easily see, how the US used their power, and in particular, mostly military power in the Middle East, in order to influence and shape the region according to their own national interest. On the one hand, by supplying arms and on the other hand by guaranteeing Saudi Arabia's and Iran's safety in case of an outside attack. In the case of Saudi Arabia, one could also see how interweaved the national interest of some petro states with a huge sovereign wealth fund is with the world economy. However, until this point (chp. 7.4) of this book, one could only see one divergence from IR or IPE realists patterns, when it came to the security of oil. That was the reaction to the Yom Kippur war. The second divergence could be the establishment of the IEA and the IEF. However, both international institutions can be explained by liberals as well as realists approaches. In the case of the protectionist measures for the oil industry in the US, one could recognise strong differences between IPE and IR realists' positions. IR and IPE liberals however, were more or less agreeing on the negativity of these measures, whereby each school was arguing according to its perspective (On the one hand only the political and on the other hand primarily the economical perspective).

Evaluating all aforementioned facts, one could say that the price, which especially the US are paying to secure oil (in particular from the Middle East), is very high due to the fact that the US is mainly following a realist approach and it could be argued that states, which are strongly relying on the market by choosing a liberal approach such as Germany or Sweden, chose more wisely. However, one could always argue that since these states are part of the 'West', they are simply taking a cheap ride under the US protection as opportunists and that there has not been a serious conflict in the world comparable to WWI or II for over 60 years, in which the advantage of controlling the flow of oil would have come into play. Furthermore, realists would always argue that it is just a matter of time until a major conflict occurs again or that, only due to the supremacy based on the control of this vital source, a conflict did not break out as the momentum of deterrence is too big.

Finally, the developments in the Arctic can be thoroughly explained by different theoretical approaches, as mentioned above, and an armed conflict seems to be, even if the Arctic oil became more and more accessible and the oil of this region streamed into the market, the centre of gravity from the supply side may also shift away from the Middle East for the 'Western' nations. However, the importance and influence of the Middle East, as shown especially in paragraph seven, will remain unchallenged, since it is the place, very unlikely. Nevertheless which can potentially drive every competitor out of the market due to its very low production costs. With these facts in mind, the political struggle in this area of the world is not likely to become less intense.

Appendix

Oil as a financial commodity

Nowadays, oil became also a financial commodity due to the fact that it is co-moving negatively with the US dollars exchange rate. Therefore, oil has become a hedge against a weaker dollar and the inflation risk (Yergin 2011: 165).[60] However, this is not without risk as the high oil prices at the end of the last decade showed. Among other reasons especially the financial markets built up a price bubble, which was putting the world economy at risk. Due to this risk Saudi Arabia reacted and increased its production in order to lower the price. However, no one in the market was eager to buy but the price kept increasing. The physical and the paper barrel did not match each other anymore. And when the price hit 130$ per barrel new car sales in the US were crashing, representing a core US industry. When the prices hit $139 per barrel the airline industry felt its back against the wall (Yergin 2011: 175-178). Together with the housing slow down the high oil price tipped the US economy into recession in 2008 (Yergin 2011: 182).

Additionally, the subsidisation of fuel in many developing countries led to massive costs explosion to keep the price 'stable'. India for example spent $21 billion on subsidies for fuel, in order to keep the prices stable at the gas stations (Yergin 2011: 183).

The 'reversed Midas touch'

The 'reversed Midas touch' is the second and even more debilitating illness of the petro-state, which is caused by excessive government spending's, irrespective of the macroeconomic situation (Yergin 2011: 107-111).

When the revenues of the government increase, it is forced by its society to increase its spending as fast as it can. More subsidies have to be handed out, more programs have to be initialised, more and big projects have to be launched and promoted and corruption also commonly spreads.

[60] In times, when central banks are flooding the monetary market with cheap money as it is happening since 2008, this phenomenon is of course only getting stronger due to the high inflation risk that is going along with this policy.

The catch of the oil industry is, however, that it is a capital-intensive one meaning that even though it is generating huge revenues it is not employing many people, which is adding further to the pressure on the state to spend revenues immediately on projects, entitlements and welfare. However, when the revenues decrease, the states dare not cut back on it's spending. The states get locked in a vicious circle of ever increasing spending. Furthermore these governments are forced by social pressure to provide very cheap oil for their citizens as an 'entitlement' for living in an oil-exporting country. This obviously leads to an inefficient and wasteful use of energy, thus reducing the supply quantities for exports and the source of the wealth of these countries. Saudi Arabia for example is using 10 per cent of its annual oil production just for electricity in their own country and of course the oil is not calculated on the world market price (Rickens 2012; Yergin 2011: 107-109, 294). Furthermore some of these countries even have to buy refined products on the world market due to a lack of refining capacity in their own countries and are therefore subsidising home fuel. However, a cut back on governments spending is putting the survival of the state/government at risk.
States in most cases try to counter the 'reversed Midas touch' by printing money (which leads to high inflation) or by foreign borrowing, which is keeping the money flow up on short term. However, when the debt and interest payments are increasing the potential of a debt crisis is rising with it (Yergin 2011: 107-109).

Nevertheless, in small countries such as the countries in the Persian Gulf the pressure of the population is not too massive and the volatile oil prices can be countered by the diversification of the national sovereign wealth funds (which is also a useful tool against the Dutch disease). Nevertheless, bigger countries such as Nigeria (Oil and Gas account for 40% of GDP), Venezuela and Russia don't have this 'blessing' of a small population in this certain context. For example Russian governmental revenues are generated by 50% from the export of oil and gas without having the luxury of being able to afford a wealth fund such as the Saudi Arabia or in this context especially to mention Norway.

OPEC members additionally have to pay attention, at least theoretically, to production quotas, meaning that they can theoretically not export more oil in order to increase their revenues. Furthermore, if the increase in production is to high, the price will decline and the problem will not be solved (Yergin 2011: 108-

111, 132). This lesson had especially to be learned by OPEC in 1998 when it decided to increase its output directly before the financial crisis hit Asia with all its might and the oil price decreased consequently dramatically (Yergin 2011: 161). The only country that is the 'in-official' supplier of stable prices in this context is Saudi Arabia, which has a policy of having always a minimum of 1.5 to 2 mbpd. as spare capacity to buffer higher demand and on the contrary it is the only country that is really cutting back supply from the market when prices tend to fall (Yergin 2011: 163). However, the social stability of these states is also in the interest of the consuming countries due to the fact that interruptions in the production can have huge impacts on the global oil market as the civil war in Libya illustrated (Yergin 2011: 293).

Posted Price

The 'posted price' was a fictitious fixed price, which was set by the seven major oil companies until 1971. It was serving as a reference point for calculating royalties and taxes, which these companies had to pay to the host countries. Furthermore, it served as a reference price for inter-company crude oil sales. By the end of the 1950s, the posted price was serving as a basis for discounts in the free market, which just had started to develop. The concept of the 'posted price' ended as a result of the Tehran Agreement in 1971 (Chalabi 2010: 296).

OPEC

Country	Region	Joined OPEC	Production in 2011 in Tb/d
Algeria	Africa	1969	15.074
Angola	Africa	2007	20.185
Ecuador	Latin America	1973	5.838
Iran	Middle East	1960	43.407
Iraq	Middle East	1960	32.040
Kuwait	Middle East	1960	30.270
Libya	Africa	1962	5.572
Nigeria	Africa	1971	25.442
Qatar	Middle East	1961	9.723
Saudi Arabia	Middle East	1960	111.589
UAE	Middle East	1967	30.127
Venezuela	Latin America	1960	28.504

Former Members:

Country	Region	Joined OPEC	Left OPEC
Gabon	Africa	1975	1995
Indonesia	East Asia	1962	2009

(OPEC 2011; OPEC 2012)

Map of US client-states in the Arab gulf

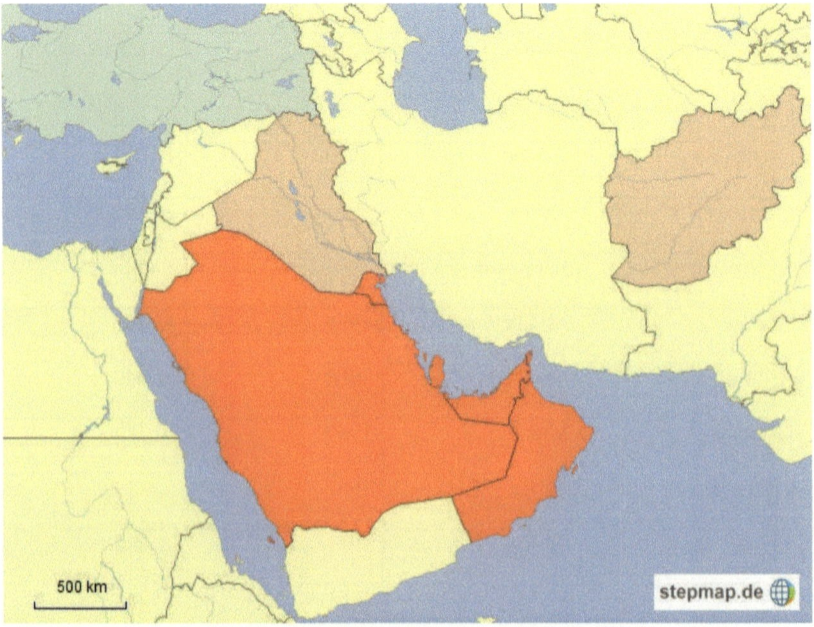

(Map created according to: O'Donnell 2007; Pierre 2011)

Orange states = are US client states

Blue states= are NATO states

Brown states = were or are recently 'occupied' by 'Western' and especially US troops

Ownership links between major IOCs and major crude-oil producing companies in the Middle East 1966 (percentage of shares held)

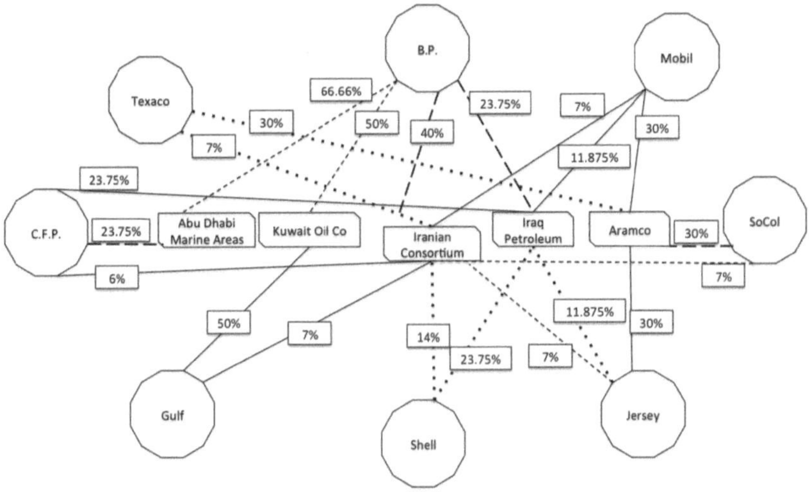

(Illustration according to Chalabi 2010: 38)

Geopolitics and oil price

(Chalabi 2010: 9)[61]

[61] The invasion of Afghanistan by the Soviet Union in 1979 is missing in this illustration. The oil price is the NPV of 2008.

The End of the Bretton Woods System

The Bretton Woods System was a system of fixed exchange rates, which was established after WWII. The system was based on the guarantee by the US to exchange if required $35 for one ounce of gold. After its establishment the exchange rates got adjusted several times in order to meet the developments of the world economy. However, on August 15 1971, this system collapsed including the fixed price for gold. This had far-reaching consequences for the value of the US Dollar. More than two years later, in October 1973 OPEC charged $4.31 US per barrel. However, if contracts would have been fixed to gold after the end of the Bretton Woods system, OPEC countries would have received $11.83 per barrel, if contracts would have been made in Yen, OPEC would have received $5.82 per barrel.

Furthermore, prior to the price increase of January 1974 the real price of oil for OPEC had been declining compared to nearly all 'Western' currencies and virtually to all other commodities in the world market. Nevertheless, from 1971 to mid 1973 the value of the dollar, the trading currency of oil, had lost about 30% of its value against the Deutschmark and the Swiss franc and more then 20% of its value to the currencies of France, Japan, Holland and Sweden. On average the value of the US dollar fell by 25% during that specific period. Furthermore, the annual inflation rate rose to more then 5 per cent on average in the early 1970 and reached roughly 9% for the decade. As a consequence prices for almost every commodity increased, not only gold: pig iron by 200%, aluminium by 165%, lead by 170%, potash by 269%, silver by 1,065% and tin by 219%.

In 1970 OPEC countries needed to sell about 10 barrel of 'black gold' in order to purchase one ounce of real gold. By the end of the Bretton Woods system OPEC would have needed 12 barrels to do the same and by mid-1973 almost 35 barrels. However, by raising the oil price in 1974 and after, the historical range of the 'gold price' for oil stayed at its ratio of about 10 to 12 barrels of oil for one ounce of gold (Hammes and Wills 2005: 501-510; Chalabi 2010: 68-69).

In short, the end of the Bretton Woods system cannot be claimed to be the only reason for the high oil price. However, it should not be forgotten in this context next to the political events that took place in the Middle East region during that time. The decline of the value of the dollar and inflation in all 'Western' countries had to be met with higher prices by oil exporting countries if they did not want to loose revenues.

Another very positive side effect of the high oil price was of course a higher demand for the US dollar as the trade currency of oil. Consequently, the higher oil price was helping to stabilise the value of the dollar (Chalabi 2010: 248).

OPEC and its role as swinging supplier

When global oil demand is growing or decreasing it has direct influence on OPEC countries due to the fact that OPEC is the world's only residual supplier of oil. In this function OPEC (and especially Saudi Arabia) is filling the gap between global oil demand and non-OPEC crude. By aiming to produce just enough oil to fill the gap between supply and demand, OPEC notionally bestows price stability to the global oil market. An overproduction would create an instant glut and downward pressure on the price; consequently if OPEC is producing less oil then needed by the market, there would be an oil shortage and the prices would rise. Although, from a historical point of view, one could argue that OPEC has not been particularly adroit in balancing its production regulation in harmony with these two magnitudes, one has to admit, that if it were not for OPEC's production restraints, price wars and market chaos would be rife as the situation in the oil market in 1986 and 1998 showed. As a consequence the production of high cost oil in areas of the North Sea or Alaska would be unthinkable (Chalabi 2010: 262-282).

Hence OPEC countries and especially Saudi Arabia are the key countries when it comes to the oil price and all its positive and negative consequences on the global energy diversity and energy costs. Countries might manage to become independent from the physical OPEC oil but not from the oil price, which is set in the Middle East.

Balance of Trade

The balance of trade is the difference between a countries imports and exports. If the net exports (exports minus imports) of a country are positive the country is selling more to the world market or a specific country than it is buying from it and vice versa (Mankiw 2004: 732-733). If a country has a negative trade balance it has to finance the difference between its imports and export via the international monetary markets, which bears all the well-known risks.

When we are having a closer look at the US and their balance of payment, it becomes visible that especially with oil producing countries the US have a negative balance of trade. For example, the nominal trade deficit between 2001 and 2011 between the US and Saudi Arabia accumulated to 221,552.7 million dollars, with the UAE to 82,256.6 million dollars, with Venezuela to 216,256.4 million dollars and with Russia to 137,941.3 million dollars, adding up 658,007 million dollars (US Census 2012).[62;63] Even Germany as a nation, which is generally characterised by a surplus in its balance of trade is showing a negative balance when it comes to energy imports in the period from 2001 to 2011 (-43.4 Million dollars) (Destatis 2012). However, it is also questionable, wether the vast amounts of money that once left the country is coming back in form of FDI and even if it does, the sovereignty is lying with the investing country and not vice versa.

The Arctic Council

In 1996 the Ottawa Declaration formally established the Arctic Council. The Arctic Council has since then been a high level intergovernmental forum to provide a means for promoting cooperation, coordination and interaction among the Arctic States. Also involved are the Arctic Indigenous communities as well as other Arctic inhabitants on common Arctic issues, especially issues of sustainable development and environmental protection in the Arctic. Permanent Member are Canada, Denmark, Finland, Iceland, Norway, Russian Federation, Sweden and the United States (Arctic-Council unknown year).

---

[62] The dollar values of the day got summarised, without adjusting them to a net present value.

[63] Within the last 26 years, the US only had a surplus with Saudi Arabia in 1985 and 1998 (both years were characterised by a crash in the oil price), with Venezuela the US had no surplus during the same time frame and with Russia only at the beginning of the trade relations in 1992 and 1993 (the balance of trade with the USSR was insignificantly small between 1985 and 1992).

Bibliography

Aust, Stefan, Richter, Claus 2011: 'Das Blut der Welt/ The World's Blood', ZDF (Public German Television). Available at: <www.zdf.de/ZDFmediathek/beitrag/video/1494904/Das-Blut-der-Welt---Teil-1; www.zdf.de/ZDFmediathek/beitrag/video/1494904/Das-Blut-der-Welt---Teil-2> [Accessed 03 June 2012].

Arctic-Council, unknown year: 'About the Arctic-Council'. [online]. Available at: < www.arctic-council.org/index.php/en/about > [Accessed 11 June 2012].

BBC News, unknown year: 'US withdrawal (1969 - 73)'. [online]. Available at: <news.bbc.co.uk/1/shared/spl/hi/asia_pac/05/vietnam_war/html/us_withdrawal.stm> [Accessed 18 June 2012].

Brauer, Birgit, 2001: 'Oil Begins Flowing Through Kazakh Pipeline'. The New York Times [online] 27th March. Available at: <www.nytimes.com/2001/03/27/business/oil-begins-flowing-through-kazakh-pipeline.html> [Accessed 17 June 2012].

Bronson, Rachel 2006: Thicker Than Oil, America's uneasy partnership with Saudi Arabia, Oxford University Press: Oxford

Chalabi, Fadhil, J. 2010: Oil Politics and Myths, Observations of an OPEC Insoder, I.B. Tauris: New York

CIA World Fact Book 2012: 'Middle East: Iran'. [online]. Available at: <www.cia.gov/library/publications/the-world-factbook/geos/ir.html> [Accessed 12 July 2012].

Copulos, Milton 2003: 'America's Achilles Heel, The Hidden Costs of Imported Oil A Strategy for Energy Independence'. The National Defense Council Foundation Washington, D.C. [online] Available at: <ndcf.dyndns.org/ndcf/energy/NDCF_Hidden_Costs_of_Imported_Oil.pdf> [Accessed 11 July 2012].

Cohn, Theodore H. 2010: Global Political Economy, Theory and Practice, 5th edition, Pearson: New York

Daily Bell unknown year: 'Friedrich August von Hayek'. [online] Available at: <www.thedailybell.com/floatWindow.cfm?id=939> [Accessed 21 June 2012].

Destatis 2012: 'Genesis-Online Datenbank'. [online] Available at: <www-gene-sis.destatis.de/genesis/online/data;jsessionid=06F371E4783217725D93A9CD34764985.tomcat_GO_2_1?operation=abruftabelleAbrufen&selectionname=51000-0003&levelindex=1&levelid=1341378540629&index=3> [Accessed 14 June 2012].

ddaymuseum.co.uk, unknown year: 'D-Day and the Battle of Normandy: Your Questions Answered'. [online]. Available at: <www.ddaymuseum.co.uk/d-day/d-day-and-the-battle-of-normandy-your-questions-answered#overlord> [Accessed 6 July 2012].

Downey, Morgan 2009: Oil 101, Wooden Table Press: New York

Economist, 2012: 'The vanishing north, What the melting of the Arctic means for trade, energy and the environment'. [online] Available at: <www.economist.com/printedition/specialreports> [Accessed 23 July 2012].

Engdahl, William 2003: A Century Of War, Anglo-American Oil Politics and the New World Order, Pluto Press: London

Fiedler, Anke 2012: 'Wirtschaft 2012: Sprudelndes Öl, starkes Wachstum/ Economy 2012: bubbling oil, strong growth'. Wirtschaftsplattform Irak [online] Available at: <www.wp-irak.de/index.php/component/content/article/40-ueberblick-a-zahlen/659-wirtschaft-2012-sprudelndes-oel-starkes-wachstum> [Accessed 25 July 2012].

Friedman, Milton, 1975, "Letter to General Pinochet on Our Return from Chile and His Reply", [online] available at: <wwww.naomiklein.org/files/resources/pdfs/friedman-pinochet-letters.pdf> [Accessed 29 July 2012].

Geology.com, unknown year: 'Largest Lake in the World'. [online] Available at: <geology.com/records/largest-lake.shtml> [Accessed 17 June 2012].

globalsecurity.org, 2012: 'Singapore Changi naval base' [online] Available at: <www.globalsecurity.org/military/facility/singapore.htm> [Accessed 18 June 2012].

Golts, Alexandr 2011: 'The Arctic: A Clash of Interests or Clash of Ambitions', *Strategic Studies Institute* [online] Available at: <www.strategicstudiesinstitute.army.mil/pdffiles/PUB1073.pdf> [Accessed 30 June 2012].

González, Ángel 2012: 'Amerika befreit sich aus der Abhängigkeit der Ölscheichs/ America is freeing itself from the dependence of the Oil Sheikhs'. The Wall Street Journal [online] 27th June. Available at: <www.wallstreetjournal.de/article/SB10001424052702303649504577492551527196854.html > [Accessed 17 July 2012].

Hammes, David, Wills Douglas 2005: Black Gold, Oil-Price Shocks of the 1970s, *The Independent Review*, v. IX, Nr. 4, Spring 2005
Holmes, James, Yoshihara, Toshi 2010: 'Understanding Asia-Pacific Sea Power'. The Diplomat [online] 21st October. Available at: <thediplomat.com/2010/10/21/understanding-asia-pacific-sea-power> [Accessed 14 June 2012].

Jackson, Robert and Sørensen, George (1999). Introduction to International Relations, Oxford: Oxford University Press

Johnston, Peter 2010: 'Arctic Energy Resources and Global Energy Security', *The Journal of Military and Strategic Studies* [online] Available at: <www.jmss.org/jmss/index.php/jmss/article/view/298/> [Accessed 29 June 2012].

Kennedy, Paul 1988: The Rise and Fall of the Great Powers: Economic Change and Military Conflict from 1500 to 2000, Random House: New York

Klare, Michael 2003: Blood and Oil, Penguin Books: London

Laruelle, Marlène 2011: 'Russian Military Presence in the High North: Projection of Power and Capacities of Action', *Strategic Studies Institute* [online] Available at: <www.strategicstudiesinstitute.army.mil/pdffiles/PUB1073.pdf> [Accessed 30 June 2012].

Lebow, Richard, N. 2010: Classical Realism, in Dunne, Tim, Kurki, Miljan, Smith, Steve 2010: International Relations Theories, 2nd edition, Oxford University Press: Oxford

Mankiw, Gregory 2004: Grundzüge der Volkswirtschaftslehre/ Macroeconomics, 3rd edition, Schäffer-Poeschel Verlag: Stuttgart

Mammadyarov, Elmar 2007: 'Azerbaijan wants regional peace and closer partnership with the West'. Deutsche Gesellscahft für Auswärtige Politik/ German Society for Foreign Affairs [online] 1st April. Available at: <ip-journal.dgap.org/en/article/20689> [Accessed 11 June 2012]

McElroy, Damien 2008: 'Georgia: Russia targets key oil pipeline with over 50 missiles'. The Telegraph [online] 10th August. Available at: <www.telegraph.co.uk/news/worldnews/europe/georgia/2534767/Georgi

a-Russia-targets-key-oil-pipeline-with-over-50-missiles.html#> [Accessed 17 June 2012].

Morgenthau, Hans J., (1978). "Politics Among Nations: The Struggle for Power and Peace", Fifth Edition, Revised (New York: A. Knopf) pp. 4-15

NavyAdvancement.com 2011: 'US Navy Bases Overseas'. [online]. Available at: <www.navyadvancement.com/military-bases/navy-bases-overseas.php> [Accessed 30 June 2012].
O'Brien, Robert, Williams, Marc 2010: Global Political Economy, 3rd edition, Palgrave MacMillan: New York

O'Donnell, Dan 2007: 'The global political economy of oil and US policy'. The University of Michigan, Centre for Middle East and North African Studies Michigan

OPEC 2011: 'Member Countries'. [online]. Available at: <www.opec.org/opec_web/en/about_us/25.htm> [Accessed 30 June 2012].

OPEC 2012: 'Market Indicators 2011'. [online]. Available at: <www.opec.org/opec_web/en/data_graphs/1993.htm> [Accessed 30 June 2012].

Parra, Francisco 2010: Oil Politics, a modern history of petroleum, I.B. Taurus: London, New York

Peaple, Andrew 2012: 'Dem Ölpreis droht der Absturz ins Bodenlose/ The Oil price is about to crash'. Die Welt [online] 25 June. Available at: <www.welt.de/wall-street-journal/article107262916/Dem-Oelpreis-droht-der-Absturz-ins-Bodenlose.html> [Accessed 25 June 2012].

Peak Oil Optimist 2005: 'Azerbaijan, Georgia, Turkey Inaugurate New Oil Pipeline'. [online]. Available at: <peakoiloptimist.blogspot.co.uk/2005/05/azerbaijan-georgia-turkey-inaugurate.html> [Accessed 15 June 2012].

Pierre, Tristan 2011: 'The Egyptian Revolution in the Context of U.S. Policy in the Middle East'. About.com [online] Available at: <middleeast.about.com/od/egypt/a/us-egyptian-revolution_2.htm> [Accessed 17 June 2012].

Ravenhill, John 2008: Global Political Economy,2nd edition, Oxford University Press: Oxford

Reynolds, David 2000: 'Britannia Overruled, British Policy and World Power in the 20th century ', 2nd edition, Pearson Education: London

Rickens, Christian 2012: 'Hat jemand was von Panzern gesagt? / Did anybody say something about tanks?'. Der Spiegel [online] 6 June. Available at: < www.spiegel.de/wirtschaft/roesler-in-saudi-arabien-panzer-heikles-thema-a-837470.html > [Accessed 6 June 2012].

Russett, Bruce 2010: Liberalism. ed. in Dunne, Tim, Kurki, Milja, Smith, Steve 2010: International Relations Theories, Discipline and Diversity. 2nd Edition, Oxford University Press: Oxford

The Lugar Energy Initiative unknown year: 'World Oil Chokepoints'. [online] Available at: <lu-gar.senate.gov/energy/security/chokepoints.cfm> [Accessed 11 June 2012].

US Census Bureau 2012: 'US Trade in Goods by Country'. [online] Available at: <www.census.gov/foreign-trade/balance/> [Accessed 1 June 2012].

U.S. Department of State, Office of the Historian unknown year: 'Mile-stones: 1969-1976: Arab-Israeli War 1973'. [online]. Available at: <history.state.gov/milestones/1969-1976/ArabIsraeliWar73> [Accessed 15 June 2012].

U.S. Energy Information Administration 2012: 'OPEC Revenues Fact Sheet'. [online]. Available at: <205.254.135.7/emeu/cabs/OPEC_Revenues/Factsheet.html> [Accessed 14 June 2012].

Wezeman, Siemon 2012: 'Military Capabilities in the Arctic', *Stockholm International Peace Research Institute* [online] Available at: <books.sipri.org/product_info?c_product_id=442> [Accessed 27 June 2012].

Yergin, Daniel 2008: The Prize, The Epic Quest for Oil, Money and Power, Simon and Schuster: New York, London

Yergin Daniel 2011: The Quest; Energy, Security, and the Remaking of the Modern World, Penguin Group: London

YI, XIAOXIONG 2009: 'The Chinese navy is going blue water'. Zanesvil-leTimesRecorder.com [online] 4th November. Available at: <www.marietta.edu/News/newspdfs/ZTR-110409.pdf> [Accessed 17 June 2012].